ENVIRONMENTAL SCIENCE, ENGINEERING AND TECHNOLOGY

WATER CONDENSATION

PROCESSES, MODELING AND CONTROL

Environmental Science, Engineering and Technology

Additional books in this series can be found on Nova's website under the Series tab.

Additional E-books in this series can be found on Nova's website under the E-books tab.

Water Resource Planning, Development and Management

Additional books in this series can be found on Nova's website under the Series tab.

Additional E-books in this series can be found on Nova's website under the E-books tab.

ENVIRONMENTAL SCIENCE, ENGINEERING AND TECHNOLOGY

WATER CONDENSATION

PROCESSES, MODELING AND CONTROL

AMIMUL AHSAN
EDITOR

Nova Science Publishers, Inc.
New York

For permission to use material from this book please contact us:
Telephone 631-231-7269; Fax 631-231-8175
Web Site: http://www.novapublishers.com

Additional color graphics may be available in the e-book version of this book.

LIBRARY OF CONGRESS CATALOGING-IN-PUBLICATION DATA

Water condensation : processes, modeling, and control / editor, Amimul Ahsan.
p. cm.
Includes index.
ISBN 978-1-61324-874-4 (softcover)
1. Water. 2. Condensation. I. Ahsan, Amimul.
QD169.W3W34 2011
536'.44--dc23
2011017624

Published by Nova Science Publishers, Inc. † New York

CONTENTS

PREFACE

This book introduces advanced water vapor condensation processes, modeling and control technologies in order to find a sustainable solution in an industrialized society. The theoretical analysis and modeling of heat and mass transfer rates produced in a condensation process is a significant issue in a design of wide range of industrial processes and devices. It includes five advanced and revised contributions which cover the laminar film-wise condensation; the condensation at nanoscale; the synthesis of heterocyclic compounds in aqueous medium; the condensation for solar still and the effect of noncondensable gases on condensation.

Chapter 1 - The analysis and prediction of heat transfer rates produced in a condensation process is an important subject in a design of wide range of industrial processes and devices. Condensation can occur in the form of continues film or in the form of discrete droplets and the way in which the condensate forms on the surface has a significant effect on the heat transfer rate. The present study concerns with the laminar film condensation phenomenon on a vertical flat plate. By using dimensionless equations and parameters, a heat transfer coefficient of the condensate has been obtained and discussed. The conclusions are that the Nusselt theory is acceptable for low Jacob numbers at all Prandtl numbers and when Jacob number increases, the results diverge from the Nusselt theory. In the next step an analytical calculation on the problem of film condensation on a vertical flat plate by taking into account the shear forces at the liquid-vapor interface has been performed. For this problem two limiting cases have been investigated in calculating the liquid film thickness and the Nusselt number. These parameters can be categorized as condensation with high interfacial shear force and with no interfacial shear force. The results of the solution show that the velocity profile has greater values in the cases with higher interfacial forces compare to the Nusselt case that no shear stress is assumed to exert upon the condensate layer. Moreover an analytical study has been done considering the problem of film condensation of assisted flowing vapor on an isothermal vertical flat plate. Some of the conclusions are that the values of the interfacial velocity are very low compared to the vapor free stream velocity but the vapor free stream velocity will result in an increase in the liquid mass flow rate. Also thinner liquid film thickness resulted from higher values of Reynolds number will enhance the heat transfer coefficient. Totally it was concluded that when an associated vapor flow is forced past a cool wall, it will drag the liquid film along, thinning it out and enhancing heat transfer. Finally in order to consider the influences of geometry on laminar condensation heat transfer rate, an analysis on forced convection film condensation on an elliptical tube has been done. The effect of Reynolds and ellipticity of ellipse on Nusselt have been discussed. Results show that increasing ellipticity and Reynolds increases the average Nusselt number.

Chapter 2 - Undersaturated water vapor in air condenses in nanoscopic gaps formed between solid surfaces (i. e. microscopic contacts between macroscopic or microscopic bodies, microscopic pores on a solid surface, nanotubes, etc) and forms small water menisci. The phenomenon is known as capillary water condensation and the curvature radii of such nanoscopic water menisci at thermodynamic equilibrium with the water vapor are determined by Kelvin equation. The atomic force microscope (AFM) is the perfect tool for investigation of the phenomenon because it senses directly the small capillary force generated by a microscopic water meniscus. Usually, in such investigations the very sharp tip of the AFM probe approaches a sample surface, get in nanoscopic contact with it, and then withdraws until the contact is broken up. During the contact, the water vapor condenses forming a nanoscopic water bridge that is broke up during the tip withdrawal. The capillary force sensed by the AFM probe during the water bridge elongation and break up is a measure of the amount of water condensed at the contact.

This chapter presents the principles of atomic force microscopy investigation of capillary water condensation at nanoscopic contacts and reports the main results obtained by the author in such investigations. It is shown that the amount of water condensed at a nanoscopic contact depends on the contact geometry, hydrophilic property of surfaces, surface contaminants, air humidity, and, in some cases, the contact time.

Chapter 3 - The widespread application of structurally diverse heterocycles in drug development has led to the exploration of new synthetic strategies. Demand for efficient methods for novel heterocyclic systems has become a powerful driving force for advances in synthetic organic chemistry. In view of the significance attached to green chemistry and its relevance to the present day problem of global warming, the development of novel, simple, cleaner synthetic protocols is attracting attention in both academic and industrial research, resulting in an ever increasing number of publications or reports on this topic. Designing environ friendly synthetic strategies in water, minimizing the use of harmful, toxic, and flammable organic solvents and hazardous reagents/catalysts, is attaining the priority over other issues. Even though water has the status of universally acceptable solvent, its utility as a solvent medium for organic transformations has severe limitations, due to the solubility problems. The present account is an attempt is to revisit briefly into several recent aspects of water medium reactions involving the synthesis of potential heterocyclic systems, which will be useful to research community.

Chapter 4 - Solar water distillation might be one of the viable options for providing drinking water for a single house or a small community in an arid, remote or coastal region. A new type of solar distillation called a Tubular Solar Still (TSS) was designed to overcome some disadvantages in the maintenance and management of a basin type still. A production model of a TSS based on a film-wise condensation theory that takes into account of thermal resistance in unsaturated humid air inside the still is presented in this chapter. The condensation coefficient due to thermal resistance is inversely proportional to the dry air pressure fraction. In the present model, the overall heat transfer coefficient between the humid air inside the still and the ambient air outside the still is used for the first time because the measurement of the ambient air temperature is easier to obtain than the inner surface temperature of the tubular cover. The analytical solution of this condensation theory could provide a good agreement with the observed production flux obtained from the field experiment and proves that the model can be used to precisely predict the production flux. A basic theoretical approach on laminar film condensation is presented first in this chapter to facilitate readers understanding before introducing the theory of film-wise condensation for a solar still.

Chapter 5 - In this chapter the authors consider the effect of a noncondensable gas on water vapor condensation. Two geometries are investigated: condensation between two parallel plane plates and condensation in a finned tube heat exchanger (cold battery). The authors start by a parametric study of condensation between two parallel plates. The model used for water vapor condensation is capable of predicting correctly the thermal performances with an error which does not exceed 7% of the measurements. The numerical results show that the mass fraction of air at the condensation surface increases with the mass fraction at inlet and there is a big concentration of the noncondensable gas at the liquid/mixture interface, which reduces the water vapor mass fraction and the condensate rate. The increase of inlet velocity leads to an increase of vapor mass fraction in the core of the mixture and the condensation surface and consequently an increase of the condensate rate. The increase of the molecular mass of the noncondensable gas leads to an increase of the water vapor mass fraction near the condensation surface(plate) and of the condensate rate. The increase in the inlet pressure causes a decrease of the air mass fraction near the wall and an increase of the water vapor mass fraction and the condensate rate. In the case of a cold battery, the air mass fraction increases more rapidly near the fin than in the middle of the channel. The maximum value of the air mass fraction is found in the recirculation zone behind the tubes. The air mass fraction around the first tube(fin side) is smaller than its middle side value. The increase in inlet velocity causes a decrease of mass fraction around the tubes and increases the condensate rate and the heat transfer coefficient. Globally the condensate rate increases with velocity.

ABOUT THE EDITOR

Dr. Amimul Ahsan was born in Netrokona, Bangladesh, 1982. He is a Senior Lecturer, Department of Civil Engineering, University Putra Malaysia, Malaysia. He received a Ph.D in Civil Engineering from the University of Fukui, Japan. He has been involved in collaboration research with many researchers and scientists in the world since 2004 and published in diverse areas of Environmental Engineering. He has published 2 books, 4 book chapters, over 18 technical papers, and is a member of the editorial board of numerous international technical journals. He has been serving on the scientific and editorial board of InTech Publishers, Austria, European Union since 2010.

Amimul Ahsan, Ph.D.
Department of Civil Engineering
Faculty of Engineering
University Putra Malaysia
Malaysia

ACKNOWLEDGMENTS

All praise be to Almighty Allah, the Creator and the Sustainer of the world, the Most Beneficent, Most Benevolent, Most Merciful, and Master of the Day of Judgment. He is Omnipresent and Omnipotent. He is the King of all kings of the world. In His hand is all good. Certainly, over all things Allah hast power.

The editor would like to express appreciation to all who have helped to prepare this book. The editor expresses the gratefulness to Carra Feagaiga, Stephanie Gonzalez, Richard Schortemeyer III, Frank Columbus and Nadya Gotsiridze-Columbus, Nova Science Publishers, Inc., NY, USA. In addition, the editor appreciatively remembers the assistance of all authors and reviewers of this book.

Gratitude is expressed to Mrs. Ahsan, Ibrahim Bin Ahsan, Mother, Father, Mother-in-Law, Father-in-Law, and Brothers and Sisters for their endless inspirations, mental supports and also necessary help whenever any difficulty.

In: Water Condensation
Editor: Amimul Ahsan

ISBN: 978-1-61324-874-4

Chapter 1

MODELING OF LAMINAR FILM-WISE CONDENSATION

J. A. Esfahani*[1] *and S. M. Modirkhazeni

Department of Mechanical Engineering, Faculty of Engineering,
Ferdowsi University of Mashhad, Mashhad, Iran

ABSTRACT

The analysis and prediction of heat transfer rates produced in a condensation process is an important subject in a design of wide range of industrial processes and devices. Condensation can occur in the form of continues film or in the form of discrete droplets and the way in which the condensate forms on the surface has a significant effect on the heat transfer rate. The present study concerns with the laminar film condensation phenomenon on a vertical flat plate. By using dimensionless equations and parameters, a heat transfer coefficient of the condensate has been obtained and discussed. The conclusions are that the Nusselt theory is acceptable for low Jacob numbers at all Prandtl numbers and when Jacob number increases, the results diverge from the Nusselt theory. In the next step an analytical calculation on the problem of film condensation on a vertical flat plate by taking into account the shear forces at the liquid-vapor interface has been performed. For this problem two limiting cases have been investigated in calculating the liquid film thickness and the Nusselt number. These parameters can be categorized as condensation with high interfacial shear force and with no interfacial shear force. The results of the solution show that

[1]E-mail: abolfazl@um.ac.ir. Tel: +98 511- 8815100 (Ext. 210) http://abolfazl.profcms.um.ac.ir.

the velocity profile has greater values in the cases with higher interfacial forces compare to the Nusselt case that no shear stress is assumed to exert upon the condensate layer. Moreover an analytical study has been done considering the problem of film condensation of assisted flowing vapor on an isothermal vertical flat plate. Some of the conclusions are that the values of the interfacial velocity are very low compared to the vapor free stream velocity but the vapor free stream velocity will result in an increase in the liquid mass flow rate. Also thinner liquid film thickness resulted from higher values of Reynolds number will enhance the heat transfer coefficient. Totally it was concluded that when an associated vapor flow is forced past a cool wall, it will drag the liquid film along, thinning it out and enhancing heat transfer. Finally in order to consider the influences of geometry on laminar condensation heat transfer rate, an analysis on forced convection film condensation on an elliptical tube has been done. The effect of Reynolds and ellipticity of ellipse on Nusselt have been discussed. Results show that increasing ellipticity and Reynolds increases the average Nusselt number.

NOMENCLATURE

A Matrix of constant coefficients

a Semi-major axis of ellipse

Ar Archimedes number

b Semi-minor axis of ellipse

C_p Specific heat of condensate at constant pressure

C_1, C_2 Function of $(\eta, \Pr)$

D_e Equivalent diameter of ellipsoid

e Ellipticity of ellipse, $\sqrt{1-(b/a)^2}$

F Dimensionless parameter, $\frac{Ra/Ja}{Re^2}$

F, G Function of $(\eta, \Pr)$

g Acceleration due to gravity

h Condensing heat transfer coefficient

h_{fg} Latent heat of condensation

h'_{fg} Latent heat of condensation corrected for condensate subcooling Jakob number

Ja Jakob number

k Thermal conductivity of condensate

L Plate length

m, n Constant powers

m'' Condensate mass flux (per unit area)

Nu Local Nusselt number, hD_e/k

$\overline{Nu}$ Average Nusselt number

p Static pressure of condensate

P Dimensionless pressure gradient parameter, $(\rho_v/\rho)Pr/Ja$

Pr Prandtl number

Q Heat transfer

Q Discrepancy vector

Re Reynolds number

Re_τ Shear Reynolds number

Ri Richardson number

T Temperature

u Velocity component in x direction

u_∞ Vapor velocity of free stream

u_e Tangential vapor velocity at the edge of boundary layer

V Vector of unknown boundary conditions

v Velocity component in y direction

x Vertical coordinate

y Horizontal coordinate

Y Vector of unknown variables

Greek Symbols

α Thermal diffusivity

α Jacobian matrix

Γ Condensate mass flow rate per unit width

Δ Average variation

δ Thickness of condensate film

δ^* Dimensionless thickness of condensate film

δ_t Thickness of thermal boundary layer

η Similarity parameter

η Dimensionless liquid film thickness

θ Non-dimensional temperature

θ Angle measured from top of the tube

μ Absolute viscosity of condensate

ν Kinematic viscosity

ρ Density of condensate

ρ_v Density of vapor

τ Interfacial vapor shear stress

ϕ Angle between the tangent to tube surface and the normal to direction of gravity

ψ Stream function

SUBSCRIPTS

c Natural scale

i, j Matrix indices

L Plate Length

n Nusselt theory Result

s Surface

sat Saturation

v Vapor

w Wall

δ Boundary layer edge

Superscripts

$-$ Indicates dimensionless

1. INTRODUCTION

Condensation is a phenomenon which occurs when a vapor at saturation temperature (T_{sat}), be in contact with a surface that is at a temperature below the T_{sat}. The latent heat of vapor is released, transferred to the surface and condensate is formed. Thus, in condensation a phase change from vapor to liquid at a surface happens.

Due to the numerical usage of condensation in a wide range of industrial devices and process, the prediction of heat transfer rates associated with this phenomenon is an essential stage. Two kinds of condensation have been observed in different procedures due to the surface condition: Drop-wise and Film-wise condensations. In drop-wise condensation, vapor condenses in the form of individual droplets while in film-wise condensation a continuous layer of condensate forms on the surface. The way that condensation happens on the surface has a significant effect on heat transfer rate. It has been observed that heat transfer coefficient in drop-wise condensation is much larger than the film-wise ones. It may be thought that because of high heat transfer rate in drop-wise condensation that results high condensation rate, this kind of condensation is more appropriate for engineering applications but it is often difficult to maintain drop-wise condition. So, design calculations of condenser, air conditioning system and other condensing devices are most of the time based on the assumption of film-wise condensation. Consequently, the focus of this chapter is on the basis of film-wise condensation.

The first reported research and experiment about condensation phenomena were from Nusselt [1]. He simplified the phenomenon by ignoring energy convection and acceleration terms and analyzed the local balance of viscous forces and the weight of the condensate. He showed that condensation heat transfer depends on local film thickness. After Nusselt, many researchers continued his study and made some improvement into it. Rohsenow formulation [2] contains the effect of energy advection but still the acceleration terms were neglected. Sparrow & Gregg [3] expanded the previous works using boundary layer theory and similarity method. In this way, they converted the governing equations to ordinary differential equations and explained the characteristic of condensation on some parameters such as Jacob and Prandtl number. They showed that if the Prandtl number is large, the influences of the inertial terms are not important. The importance of such results has been well known and has been documented in Merte [4], extended by Koh [5], Chen [6], and Tanasawa [7]. Moreover, similarity solutions of the two-phase boundary layer equations in film condensation on a flat surface have been obtained by Koh [8] for the case of forced convection. Experimental results on film condensation have been correlated by Chen et al. [9]. Mendez et al. [10] studied the conjugate heat transfer condensation process of saturated vapor associated with the effect of longitudinal heat conduction in the thermal thick wall regime. They have used perturbation methods and the boundary layer description. Recently, Xu et al. [11] have presented the Analytical approximation for laminar film condensation of saturated stream on an isothermal vertical plate. They have reduced the boundary layer equations of momentum and thermal energy to two ordinary differential equations by means of a set of similarity transformations. The problem has been then solved analytically using the homotopy analysis method. Furthermore, laminar film-wise condensation on a vertical surface has been studied by Fujii and Uehara [12]. Two-phase boundary layer equations of laminar film-wise condensation have been solved by them with an approximate method following the Jacob work. Also, Poots and Miles [13] have done an attempt on study the effects of variable physical properties on laminar film condensation of saturated steam on a vertical flat plate.

The above works consider film condensation without the effects of the shear drag at the liquid vapor interface, while limited analysis has been conducted on discussing these effects. Koh et al. [14] obtained an exact boundary layer solution, while the shear forces at the liquid-vapor interface were taken into account. Sugawara et al. [15] worked on the condensation of vapor flowing normal to a horizontal pipe and evaluated the shear stress at the condensate surface as in the case of single-phase flow and neglected the effect of condensation. Shekriladze and Gomelauri [16] used an opposite extreme (infinite condensation rate) to approximate the surface shear stress. They have considered forced convection film condensation from a vapor flowing downward to a horizontal circular tube, and obtained numerical solutions by utilizing the asymptotic shear stress at the interface, including the effects of gravity force and velocity of vapor, without accounting for the pressure gradient term around a circular tube. Hsu and Yang [17] and Rose [18] modified the Shekriladze and Gomelauri model and considered the pressure gradient effect by using the potential flow theory. Also, Yang and Chen [19] focused on the role of surface tension and ellipticity in laminar film condensation on a horizontal elliptical tube. Furthermore, Rose [20] has studied condensation heat transfer fundamentals. He has presented an outline and discussion of those aspects of condensation heat transfer theory which are relatively well unnderstood. For example the complications arising in forced convection condensation, where the condensate film is significantly affected by shear stress, have been outlined by him. Also, Rose [21] has tried an attempt to consider the problem of surface tension effects and enhancement of condensation heat transfer. His attention is focused on methods of enhancement of heat-transfer in which surface tension plays an important role. Moreover, Kibboua and Azzi [22] have studied laminar film condensation on an elliptical tube embedded in porous media. They have determined the effect of vapor shear on condensation by simultaneous solution of the two phase vapor boundary layer and condensate film momentum equations. More recently, Chang and Yeh [23] have performed a theoretical investigation to consider the effects of uniform suction and surface tension on laminar film-wise condensation on a horizontal elliptical tube in a porous medium. The theoretical results presented by them, show that the heat transfers can be increased by applying a suction effect at the wall. Furthermore, it has been shown by them that the surface tension force has a negligible effect on the mean Nusselt number.

As already mentioned, investigating condensation process on the various geometry except flat plate has attracted the attention of researchers. Film condensation heat transfer of pure vapor flowing onto a body such as a plate, cylinder and sphere has been studied by many researchers, like Winkler et al. [24] and Yang and Hsu [25]. Moreover, Al-Jarrah et al [26] have investigated numerically the problem of a two dimensional laminar liquid film which condenses on a vertical micro-channel. A liquid film thickness, condensation mass flux flow and the variation of the velocity through the liquid thickness have been determined by modified Navier–Stockes and energy equations. Also, the effects of some parameters including slipping in temperature and velocity on the liquid film thickness, condensation mass flow rate, and velocity have been studied by them. Honda and Fujii [27] investigated numerically the condensation of a flowing vapor on a horizontal tube and analyzed it as a conjugate heat transfer problem. By considering the effects of convection, Memory et al. [28, 29] and Michael et al. [30] investigated the forced convection laminar condensation on a horizontal tube in experimental data and theoretical analysis. In addition, several researchers such as Yang and Hsu [31] and Yang and Chen [32], Ali and McDonald [33], Karimi [34] and Memory et al. [35] worked on the problem of the condensation heat transfer on cylinders of elliptical profiles with major axes aligned with gravity and developed a simple mathematical model for the study of convection film condensation with downward flowing vapor onto a horizontal tube. Also, Chang and Yeh [36] have investigated numerically the problem of steady film-wise condensation flow over the outer surface of a horizontal elliptical tube with wall suction effects within a stationary saturated vapor. To facilitate the analysis, an effective suction function has been introduced by them to represent the effect of the wall suction on the thickness of the liquid film such that the local condensate film thickness and local Nusselt number can be derived using a simple numerical shooting method. They have resulted that the tube ellipticity and the wall suction effect have a significant influence on the mean Nusselt number.

To add more, Dhir and Linhard [37] presented a solution for laminar film condensation on a two-dimensional isothermal surface. Cess [38] has studied laminar film condensation on a flat plate in the absence of body force while Jacobs [39] has worked on an integral treatment of combined body force and forced convection in laminar film condensation. Enhancing the Nusselt theory, Churchill [40] investigated the effects of the inertia and the heat capacity of the condensate on the vapor drag and the curvature of the surface. Also, Mayhew et al. [41] has analyzed the effect of vapor drag on laminar film condensation on a vertical surface. Due to the study the independent temperature physical properties in film condensation, Voskresenskiy [42] and Labuntsov [43] have made relatively simple modifications for variable thermo-physical properties. Furthermore, Brouwers [44] has studied the problem of film condensation on non-isothermal vertical plates. He has done an analytical study, presenting the condensation of a pure saturated vapor on a cooled channel plate, including the interaction between the cooling liquid, the condensate, and the vapor. The results for condensation heat transfer have been obtained for a broad range of the characteristic dimensionless numbers of the process. Moreover, Shang and Adamek [45] study on laminar film of saturated steam on a vertical flat plate for consideration of various physical factors including variable thermo-physical properties, except for surface tension at the liquid-vapor interface. Karabelas and Markatos [46] presented an investigation of water vapor's condensation to liquid substance in highly convective flow conditions. To achieve their goal, they have chosen airfoil geometry to demonstrate the applicability of the model developed. Their attention was focused on the dispersion of the condensation cloud and on the areas of mass transfer interaction between liquid droplets and humid air. In addition, Dengliang et al. [47] has tried to find a convenient and practical procedure for calculating heat transfer of laminar film condensation on a vertical fluted tube. More Recently, Baojin et al. [48] have employed visual experiments to investigate heat transfer characteristics of steam on vertical titanium plates with and without surface modifications for different surface energies.

To consider the effect of non-condensable gasses, Monning and Numrich [49] have focused on condensation of vapors of immiscible liquids in the presence of a non-condensable gas. They have stated that during partial condensation of vapors from a mixture with non-condensing gas, the gas content causes a resistance to heat and mass transfer and if the two vapor components are immiscible in the liquid state, at each composition except from the heteroazeotrope, one vapor component will additionally behave as a non-condensing gas. Also, Studying of free convection film condensation in the presence of non-condensable gases has been done by Al-Diwany and Rose [50]. Minkowycz and Sparrow [51], Denny and Jusionis [52], Denny et al. [53], and Denny and south [54] have considered the effects of non-condensable gas too. More recently, theoretical modeling of condensation of steam outside different vertical geometries (tube, flat plates) in the presence of non-condensable gasses likes air and helium has been done by Ganguli et al. [55]. The effects of influencing variables such as non-condensable gas mass fraction, operating pressure and the degree of subcooling have been numerically studied by them. Moreover, Chen and Lin [56] have considered the laminar film condensation from a downward-flowing steam-air mixture onto a horizontal circular tube. They have developed a numerical model for this problem and concluded that the local Nusselt number and liquid film thickness increase as both the non-condensable air mass fraction and the tube temperature decreases. Furthermore, Chen and Lin [57] have presented a numerical method to study the turbulent film condensation in the presence of non-condensable gases over a horizontal tube. The numerical results demonstrated that a very small bulk concentration of non-condensable gas reduces the heat transfer and film thickness. Condensation on the containment in the presence of non-condensables has received attention by nuclear scientists and engineers in the last two decades. Due to this reason, a review on condensation on containment structures has been done by Rosa et al. [58]. They have reviewed the specific physical phenomena that are involved in condensation and have discussed how these phenomena have been considered in the different available models.

Adeyinka and Naterer [59] investigated the physical significance of entropy generation in plate film condensation. Using some simplifications, their results for optimized entropy generation and plate size are expressed in terms of a duty parameter. After them, Yang et al. [60] have considered the problem of Thermodynamic optimization of free convection film condensation on a horizontal elliptical tube with variable wall temperature. By using an entropy generation minimization technique, they have studied the thermo-dynamic losses caused by heat transfer and flow friction for the laminar film condensation on a non-isothermal horizontal elliptical tube. Dung and Yang [61], Li and Yang [62], Yang et al. [63], Li and Yang [64], and Chen et al. [65] have also attempted to study the entropy generation of convection film condensation on horizontal circular and elliptical tube. In another study of entropy generation, Esfahani and Koohi [66] studied the problem of film-wise condensation on a flat plate by considering shear forces at the liquid-vapor interface. Moreover, Entropy generation analysis in film condensation on an elliptical tube with interfacial shear stress at high Bond numbers has been done by Esfahani and Koohi [67]. They have investigated two limiting cases Based on the interfacial shear and the gravitational forces applied to the liquid film at high Bond numbers. These cases have been categorized as condensation with high gravitational force compared to the interfacial one and condensation with high shear force compared to the gravitational one. Also, an analysis of the mechanism of entropy generation on an elliptical tube encountered in forced convection film condensation heat transfer has been studied by Esfahani and Modirkhazeni [68, 69]. A complete thermodynamic analysis, including first and second law has been done by them.

Besides this comprehensive review on external film condensation, we try to restudy the efforts of Esfahani and Ziaei-Rad [70, 71] about investigating an analysis of laminar film condensation on a vertical plate. We will try to obtain a heat transfer coefficient of the condensate film for a wide range of physical properties by using dimensionless equations and parameters. In the next step the analytical calculation on the problem of film condensation on a vertical flat plate by taking into account the shear forces at the liquid-vapor interface like which is done by Esfahani and Koohi-Fayegh [72] will be studied. Moreover the details of the problem of film condensation of assisted flowing vapor on an isothermal vertical flat plate, with assistance of the effort of Esfahani and Koohi-Fayegh [73] will be defined. At last, in order to consider the influence of geometry, the problem of forced convection film condensation on an elliptical tube will be studied and the effect of ellipticity of ellipse and Reynolds on Nusselt number will be discussed.

2. Nusselt Theory for Laminar Film Condensation on a Vertical Plate

As in this chapter we try to study and also modify Nusselt theory by changing some of the assumption that has been used in this theory, it is better to commence our perusal by introducing the Nusselt theory.

The situation of Nusselt theory for laminar film condensation on a vertical plate is shown in figure 1 in which film condensation of saturated vapor at T_{sat}, has been formed on the cold isothermal vertical plate which is held at T_w. This figure confirms that there may be several complicating features in the process of film condensation. The film forms at the top of the plate and flows downward under the influence of gravity. The thickness of the condensate film, δ, and the condensate mass flow rate increase as the condensate film runs downward. Then, there is heat transfer from the interface of liquid-vapor through the film to the

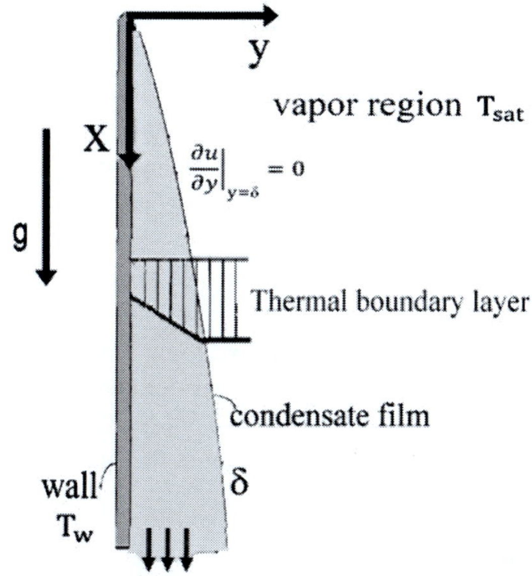

Figure 1. Film condensation on a vertical isothermal surface with assumptions of Nusselt analysis.

surface, which is at $T_w < T_{sat}$. It should be noted that in some cases the vapor may be superheated ($T_{v,\infty} > T_{sat}$) or there may be non-condensable gases in the vapor mixture.

2.1. Nusselt Assumptions

The analysis of steady film condensation on an isothermal surface can be done by using an approximate analysis of Nusselt [74]. He has applied the following assumption to the problem of film condensation:

1. The flow in the liquid film is laminar and the condensed liquid has constant properties.
2. The vapor reservoir is at rest and contains pure vapor which is everywhere at the saturation temperature, T_{sat}. With no temperature gradient in the vapor, heat transfer to the liquid-vapor interface can occur only by condensation at the interface and not by conduction from the vapor.

3. Momentum changes within the film are negligible. In writing the momentum equation for the film, the viscous, gravitational, and pressure forces acting on the film are assumed to be far greater than the rate of change of liquid momentum within the film.
4. The temperature of the liquid-vapor interface is the saturation temperature, T_{sat}, corresponding to the vapor reservoir pressure.
5. The viscous shear force of the vapor on the film at y=δ is negligible.

$$\partial u/\partial y|_{y=\delta} = 0$$

with this assumption and the foregoing assumption of a uniform vapor temperature, there is no need to consider the velocity or thermal boundary layers in vapor.
6. Pressure changes in y-direction across the film are negligible.

2.2. Control Volume

The analysis of condensation phenomenon is based on the application of the conservation of mass, momentum, and energy principles and these conservation principles should be applied to a control volume which is selected from the liquid film. Because two-dimensional flow is being assumed in the following analysis, a control volume with unit width will be considered. A sample of a control volume has been shown in Figure 2.

2.3. Subcooling Effect

When the film condensate forms on the surface, the latent heat transferred to the liquid film. The latent heat plus the sensible heat along with the cooling of the liquid film below the saturation temperature is then transferred to the wall. Most of the time, the amount of sensible heat due to the subcooling (cooling of the liquid below the saturation temperature) is negligible compared to the released latent heat. So, in order to simplify the analysis, the effect of subcooling on energy balance will be neglected. This assumption causes a linear temperature profile in the condensate film as it is shown in figure 1.

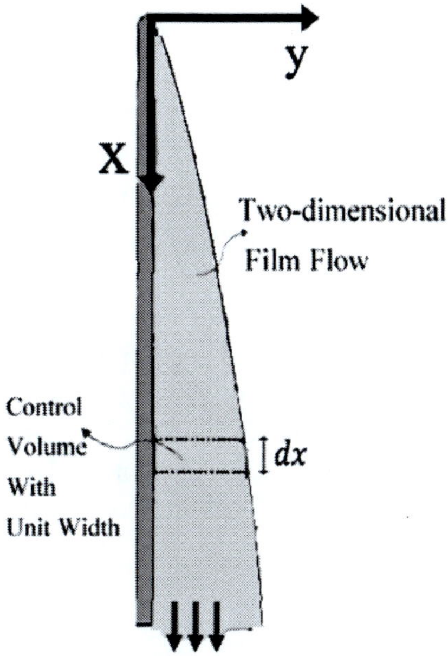

Figure 2. Type of control volume across condensed film Nusselt analysis.

2.4. Latent Heat

The latent heat, h_{fg}, produced by the phase change from vapor to liquid at interface. Nusselt [1] and Rohsenow [2] showed that if the subcooling effect has not been neglected and by considering the thermal advection effects, latent heat can be modified by adding a term to h_{fg}. So, Rohsenow recommended altering the latent heat as follow:

$$h'_{fg} = h_{fg}(1 + 0.68\,Ja) \tag{1}$$

where Ja is Jacob number which will be defined in the next sections. It should be noted that if this modification is not needed, h_{fg} itself will be used. Furthermore, Sadasivan and Lienhard [75] have investigated that the modified latent heat does not depend on the Prandtl number.

According to the following assumptions and by using the conservation principles of mass, momentum, and energy, and considering related boundary conditions, also by applying a balance of energy to a differential element of liquid condensate, the local and average heat transfer coefficients and Nusselt number can be obtained which are necessary for thermal analysis of condensation on a vertical flat plate.

3. Laminar Film Condensation on a Vertical Plate without Interfacial Shear Stress

Since most of the prior researches are about determining the basic characteristics of condensate film, in this section, the main attempt is to develop to the work of Sparrow and Gregg [3] focusing on more details using dimensionless equations and parameters [70,71]. So, with the help of using some scales and developing a computer code, the numerical values will be calculated and the results for a wide range of physical properties will be investigated.

3.1. Physical Model

The physical problem which has been shown in figure 3 contains a vertical flat plate that has been fixed in a body of pure vapor.

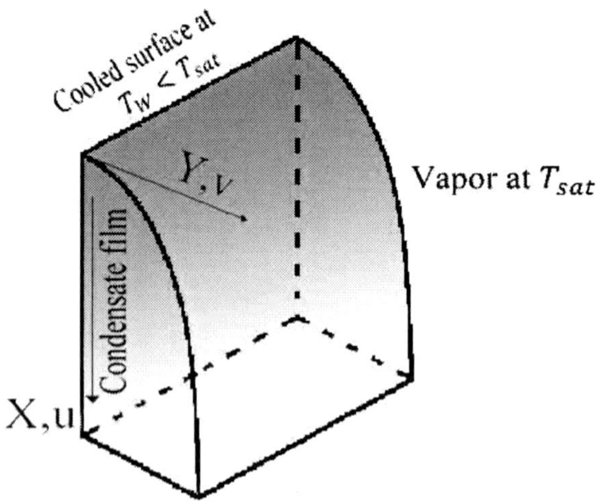

Figure 3. Physical model and related coordinate system.

The plate surface is at T_w and the saturated vapor is at $T_{sat} > T_w$, thus a continuous laminar film of condensate can forms on the plate and runs downward. The assumptions are that effect of vapor velocity and viscous dissipation can be neglected. Also it was assumed that there is no shear stress on the boundary layer edge and it is at saturated temperature.

3.2. Mathematical Model

The governing equations for this physical problem in the mentioned coordinate system can be written as:

$$\frac{\partial u}{\partial x} + \frac{\partial v}{\partial y} = 0 \tag{2}$$

$$\rho\left(u\frac{\partial u}{\partial x} + v\frac{\partial u}{\partial y}\right) = g(\rho - \rho_v) + \mu\frac{\partial^2 u}{\partial y^2} \tag{3}$$

$$\rho C_p\left(u\frac{\partial T}{\partial x} + v\frac{\partial T}{\partial y}\right) = k\frac{\partial^2 T}{\partial y^2} \tag{4}$$

where ρ_v is the vapor density, and other properties are those of condensate. Shear stress on the plate surface can be defined as:

$$\rho u\frac{\partial}{\partial x}\int_0^{\delta} u dy = -\mu \left.\frac{\partial u}{\partial y}\right|_{y=0}$$

The related boundary conditions can be considered as:

$$\text{At } y = 0\begin{cases} u = 0 \\ v = 0 \\ T = T_s \end{cases} \quad ; \quad \text{at } y = \delta\begin{cases} \frac{\partial u}{\partial y} = 0 \\ T = T_{sat} \end{cases} \tag{5}$$

The thickness of the condensate layer, δ, which is a function of x, will be determined later.

We now that there are advantages in using dimensionless form of the equations. In order to change the governing equations into dimensionless form, the dimensionless parameters have been defined as follow:

$$\bar{u} = \frac{u}{u_c}; \quad \bar{v} = \frac{v}{v_c}; \quad \bar{T} = \frac{T - T_{sat}}{T_s - T_{sat}}; \quad \bar{y} = \frac{y}{\delta}; \quad \bar{x} = \frac{x}{L} \tag{6}$$

Subscript c signifies for natural scales. Additionally, an energy balance can be written as:

$$\alpha C_p \frac{T_{sat} - T_s}{\delta} = v_c h_{fg} \tag{7}$$

This balance is between the latent heat released at the interface through condensation and the local heat flux conducted through the liquid layer to the wall surface. Since the liquid temperature is linear, Fourier's law has been used to express the surface heat flux. By using Jacob number which is defined as:

$$Ja = C_p \frac{T_{sat} - T_s}{h_{fg}} \tag{8}$$

Also, by substituting from equations (6) and (7) into governing equations (2), (3), and (4), dimensionless forms of conservation equations can be gained as:

$$\frac{\partial \bar{u}}{\partial \bar{x}} + \frac{\partial \bar{v}}{\partial \bar{y}} = 0 \tag{9}$$

$$\frac{Ja}{Pr}\left(\bar{u}\frac{\partial \bar{u}}{\partial \bar{x}} + \bar{v}\frac{\partial \bar{u}}{\partial \bar{y}}\right) = 1 + \frac{\partial^2 \bar{u}}{\partial \bar{y}^2} \tag{10}$$

$$Ja\left(\bar{u}\frac{\partial \bar{T}}{\partial \bar{x}} + \bar{v}\frac{\partial \bar{T}}{\partial \bar{y}}\right) = \frac{\partial^2 \bar{T}}{\partial \bar{y}^2} \tag{11}$$

Also we can gain another form of conservation equations Introducing stream function, ψ, which is defined as:

$$u = \frac{\partial \psi}{\partial y}, \quad v = -\frac{\partial \psi}{\partial x}$$

So, the governing equations (2), (3), and (4) have been obtained as follows:

$$\left(\frac{\partial \psi}{\partial y}\frac{\partial^2 \psi}{\partial x \partial y} - \frac{\partial \psi}{\partial x}\frac{\partial^2 \psi}{\partial y^2}\right) = \frac{g(\rho - \rho_v)}{\rho} + \nu \frac{\partial^3 \psi}{\partial y^3} \tag{12}$$

$$\left(\frac{\partial \psi}{\partial y}\frac{\partial T}{\partial x} - \frac{\partial \psi}{\partial x}\frac{\partial T}{\partial y}\right) = \alpha \frac{\partial^2 T}{\partial y^2} \tag{13}$$

The above equations can be reduced to ordinary differential equations by using similarity transformations. In order to apply similarity to the problem, the following scales have been defined as:

$$u \approx \alpha \left(\frac{k\nu}{gC_p(\rho - \rho_v)x^3}\right)^{-\frac{1}{2}} G(\eta, Pr) \ ; \ \delta_t \approx \left(\frac{k\nu}{gC_p(\rho - \rho_v)x}\right)^{\frac{1}{4}} \tag{14}$$

and the similarity parameters have been introduced as:

$$\eta = \frac{y}{\delta_t} = c\, y\, x^{-\frac{1}{4}} \ ; \quad c = \left[\frac{gC_p(\rho - \rho_v)}{4k\nu}\right]^{\frac{1}{4}} \tag{15}$$

where, δ_t is the thickness of the thermal boundary layer. The stream function has been gained as:

$$\psi = 4\alpha c x^{\frac{3}{4}} F(\eta, Pr) \tag{16}$$

Dimensionless temperature θ, has been defined as:

$$\theta(\eta) = \frac{T - T_{sat}}{T_s - T_{sat}} \tag{17}$$

Substituting the similarity parameters in equations (12) and (13), ordinary differential form of governing equations has been obtained:

$$\dddot{F} + \frac{1}{Pr}\left[3F\ddot{F} - 2\dot{F}^2\right] + 1 = 0 \tag{18}$$

$$\ddot{\theta} + 3F\dot{\theta} = 0 \tag{19}$$

and the new boundary conditions have been changed as:

$$\text{at } \eta = 0 \begin{cases} F = 0 \\ \dot{F} = 0 \\ \theta = 1 \end{cases} ; \quad \text{at } \eta = \eta_\delta \begin{cases} \ddot{F} = 0 \\ \theta = 0 \end{cases} \tag{20}$$

where η_δ is the value of η, at interface of phases $y = \delta$. For determining the value of η_δ, an overall balance of energy on condensate film should be written:

$$\underbrace{\int_0^X k\left(\frac{\partial T}{\partial y}\right)_{y=0} dx}_{\substack{\text{Heat conduction} \\ \text{from condensate} \\ \text{to the plate}}} = \underbrace{\int_0^\delta \rho u h_{fg} dy}_{\substack{\text{Energy liberated} \\ \text{as latent heat}}} + \underbrace{\int_0^\delta \rho u C_p (T_{sat} - T) dy}_{\substack{\text{Energy liberated} \\ \text{as Sub-cooling of} \\ \text{the condensate}}} \tag{21}$$

In terms of similarity variables, equation (21) can be re-written as:

$$Ja = C_p \frac{T_{sat} - T_s}{h_{fg}} = -3 \frac{F(\eta_\delta)}{\dot{\theta}(\eta_\delta)} \tag{22}$$

$\dot{\theta}(\eta_\delta)$ and $F(\eta_\delta)$ are the values of $d\theta/d\eta$ and F at $\eta = \eta_\delta$, respectively [3].

The behavior of film condensation can be analyzed by solving the equations (18) and (19) and also (22) with respect to boundary conditions in equation (20). It should be noted that two out of five boundary conditions are defined at $\eta = \eta_\delta$ and the three other at origin.

3.3. Nusselt Number

In analyzing every heat transfer problem, heat transfer coefficient and Nusselt number are important parameters that need to be considered. For the problem of condensation on a vertical plate, local Nusselt number can be defined as:

$$Nu_x = \frac{h_x x}{k} = \frac{-k\left(\frac{\partial T}{\partial y}\right)_{y=0}}{T_{sat} - T_s} \frac{x}{k} = x \left[\frac{\partial \theta}{\partial \eta} \frac{\partial \eta}{\partial y}\right]_{\eta=0} \tag{23}$$

Therefore:

$$Nu_x = \left[\frac{gC_p(\rho - \rho_v)}{4\nu k}x^3\right]^{\frac{1}{4}} \left(\frac{d\theta}{d\eta}\right)_{\eta=0} \tag{24}$$

The balance of energy which was introduced before implies that [77, 78]:

$$\frac{\rho h_{fg} u_c \delta^2}{k(T_{sat} - T_s)L} = O\,(1) \tag{25}$$

where u_c and L are the reference speed and plate length respectively. Similarly, the balance of the viscous and Archimedean buoyancy forces in the equation of motion requires that:

$$\frac{\rho \nu u_c}{\delta^2 g(\rho - \rho_v)} = O\,(1) \tag{26}$$

By using relations above and the rate of heat transfer which is defined as:

$$\dot{Q} \approx \rho u_c \delta h_{fg} \approx \frac{k\theta_c L}{\delta} \tag{27}$$

The order of Nusselt number can be gained as:

$$Nu_L = O\left(\frac{\mathrm{Pr}\, Ar_L}{Ja}\right)^{\frac{1}{4}} \tag{28}$$

where Ar_L is Archimedes number that is:

$$Ar_L = \frac{gL^3(\rho - \rho_v)}{\rho \nu^2} \tag{29}$$

The Archimedes number is similar to the Rayleigh number in free convection which is a representative of the relative magnitude of the buoyancy and viscous forces in the fluid.

Finally, the Nusselt number can be expressed as:

$$Nu_L\left(\frac{4Ja}{\mathrm{Pr}\, Ar_L}\right)^{\frac{1}{4}} = C_1(Ja, Pr) \tag{30}$$

On the other hand, using equation (24), it can be written that:

$$\left(\frac{d\theta}{d\eta}\right)_{\eta=0} = C_1(Ja, Pr)\, Ja^{-\frac{1}{4}} \tag{31}$$

Thus, this parameter has an important role in explaining the difference of heat transfer rate for high Jacob number with the Nusselt Theory. It should be noted that for $\left(\frac{d\theta}{d\eta}\right)_{\eta=0} = 1$, results tend to Nusselt theory, i.e. a linear temperature profile.

On the other hand, some results will be found on plotting the parameter $\frac{\overline{Nu}}{\overline{Nu_n}}$ versus $\mathrm{Pr}^{\mathrm{n}}\, Ja^{m}$ in various values of m and n, where:

$$\frac{\overline{Nu}}{\overline{Nu_n}} = \frac{\overline{Nu_L}}{0.943\left[\frac{g(\rho-\rho_v)L^3 h_{fg}}{\nu k(T_{sat}-T_s)}\right]^{\frac{1}{4}}} = \left(0.68+\frac{1}{Ja}\right)^{-\frac{1}{4}}\left(\frac{d\theta}{d\eta}\right)_{\eta=0} \tag{32}$$

In this relation, $\overline{Nu_n}$ is the original relation presented by Nusselt theory [1].

3.4. Method of Solution

To solve the equations, equations (18) and (19) with boundary condition (20) should be transferred to a set of ODE in the form of:

$$\dot{Y} = AY \tag{33}$$

and:

$$Y^T = \{F\ \dot{F}\ \ddot{F}\ \theta\ \dot{\theta}\},\quad Y^T(0) = \{0\ 0\ \ddot{F}(0)\ 1\ \dot{\theta}(0)\},$$
$$Y^T(1) = \{F(1)\ \dot{F}(1)\ 0\ 0\ \dot{\theta}(1)\} \tag{34}$$

where Y is a vector of five by one and A is a square matrix.

The initial conditions of the governing O.D.E. equations (18) and (19) are not entirely presented at one point. Thus, the problem to be solved is a two point's boundary value problem [70, 71]. There are two distinct classes of numerical methods for solving such problems [76]. Shooting method has been used to solve equations (18) and (19). This method is based on guessing two values for the dependent variables at beginning of domain. So the ordinary differential equations are solvable with an initial value method such as Runge-Kutta. But it should be noted that the adjustment of the free parameters at the starting point that zeros the discrepancies at the other boundary point, always should be checked. So, a multidimensional root finding problem has been solved to find the needed adjustment.

At the starting point, zero, five starting values Yi existed that are needed to be specified, but subject to three conditions. Therefore, there are two freely specifiable starting values. It can be imagined that these freely specifiable values are the components of a vector V that lives in a vector space of dimension two. The components of V might be exactly the values of certain "free" components of Y, with the other components of Y determined by the boundary conditions [70, 71]. Giving a particular V, a particular Yi (0) is thus generated. It can then be turned into a Yi (1) by integrating the ordinary differential equations to 1 as an initial value problem. Now, at 1, a discrepancy vector Q of dimension two can be defined, whose components measure how far these are from satisfying the two boundary conditions at 1. A vector value of V is found, which makes zero the vector value of Q. This is iteratively done by computation in set of linear equations:

$$[\alpha]\delta V = -Q \tag{35}$$

and then adding the correction back,

$$V^{new} = V^{old} + \delta V \tag{36}$$

In (35), the matrix $[\alpha]$ is a two by two matrix and has components given by:

$$[\alpha]_{ij} = \frac{\partial Q_i}{\partial V_j} \tag{37}$$

These partial derivatives can be computed numerically. For this idea, make a small change on Y_i vector and compute vector Q again. Then, there is:

$$\frac{\partial Q_i}{\partial V_j} \approx \frac{Q_i(V_1, V_2 + \Delta V_2) - Q_i(V_1, V_2)}{\partial V_j} \tag{38}$$

A code has been used to solve the governing O.D.E. equations (18) and (19) with present conditions. In this code two boundaries on zero has been guessed and then corrected these guesses as discussed above. An algorithm of this code has been shown in figure 4.

3.5. Results and Discussion

From non-dimensional equations, it is clear that for very small Jacob number, the temperature profile is almost linear and such a result is already presented in Nusselt theory [1]. However, as Jacob number increases, the linear approximation of the temperature profile is not true anymore. Figure 4 presented the temperature profile as a function of Jacob number. It can be seen that the trend of the temperature profile for low Jacob numbers is linear. In addition, from non-dimensional equations, it can be concluded that for viscous fluid, i.e. $Pr>>1$ the velocity profile is a parabola.

In Figure 5, $\overline{Nu_n}$ is presented for $m=1$ and $n=1/3$ where the power of 1/3 is like forced convection for high Pr. The results for $Pr=10$ and 100 coincided. Furthermore, for $m=1$, variation of n has no effect on the results, but for $m=-1$, by increasing n from 1/7 to 1, the tail-end of lines merge together.

Figure 6 depicts relative Nusselt number for different values of *Ja* and *Pr*. It can be seen that the Nusselt theory is acceptable for low Jacob numbers at all Prandtl numbers. The results have been also compared with similarity method of Sparrow and Gregg [3] presented for low Pr numbers. It can be observed from the figure that when Jacob number increases, the results diverge from the Nusselt theory. These divergences are from level one downward at low *Pr* and upward for high *Pr*. We can conclude a resulting profile at $Pr=100$ which can be used for $Pr/Ja<0.1$.

A plot of local heat transfer is also given in figure 7 for Pr=1. The agreement between these results and those of Sparrow and Gregg [3] is excellent. Their calculation technique, utilizing the similarity method, is thus shown to be completely satisfactory for Pr=1.

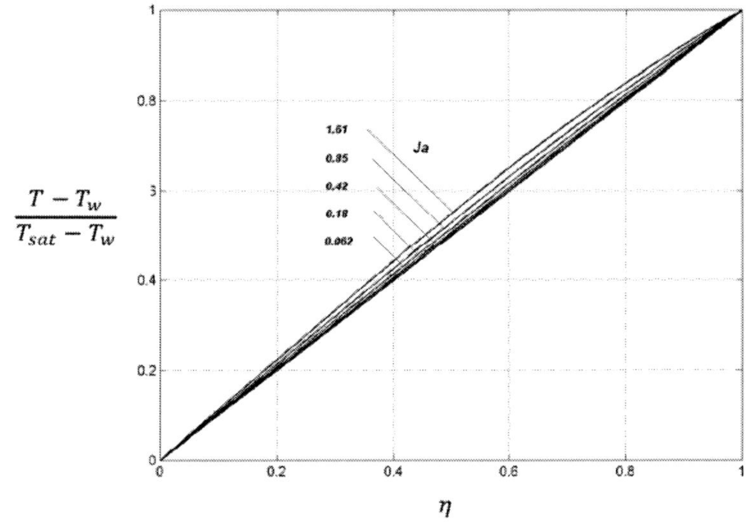

Figure 4. Temperature profile for various *Ja* numbers at *Pr=1.*

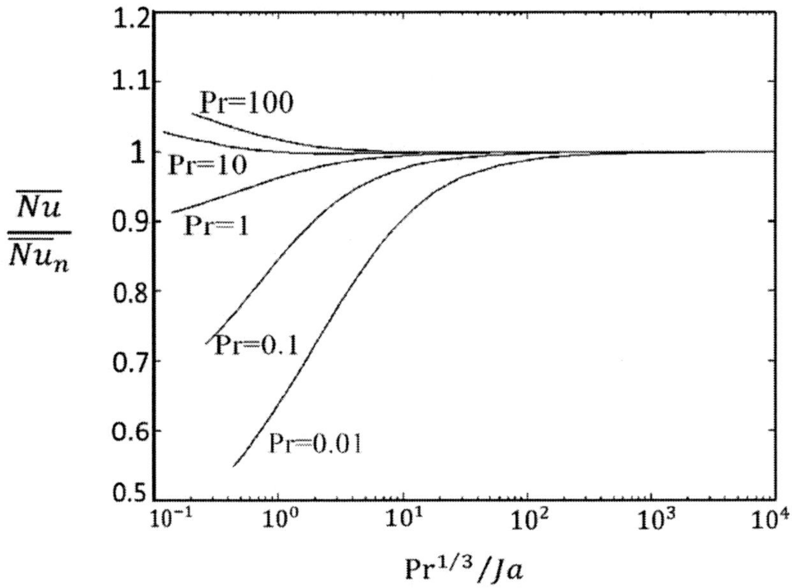

Figure 5. Variation of $\frac{\overline{Nu}}{\overline{Nu_n}}$ versus ($\mathrm{Pr}^{1/3}/Ja$) for different *Pr* Numbers.

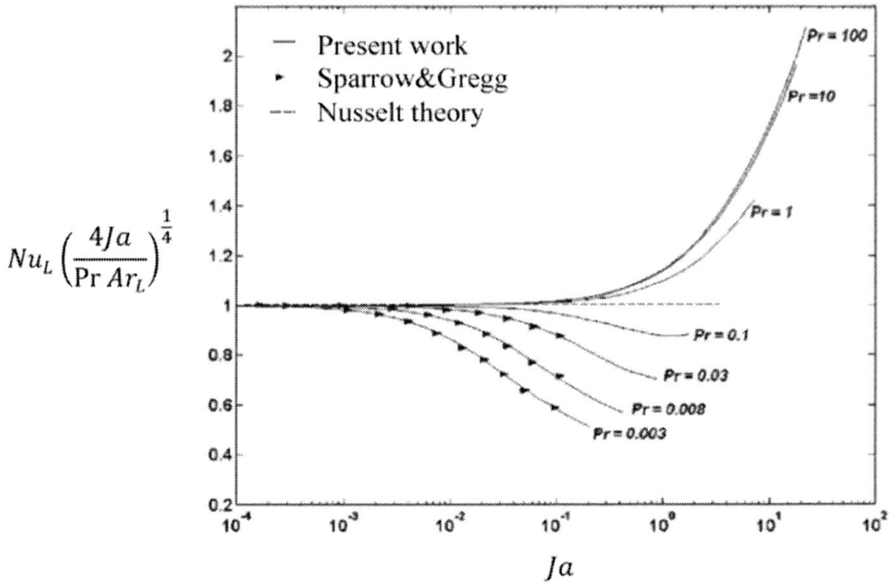

Figure 6. Variation of average Nusselt versus *Ja* number for different *Pr* numbers.

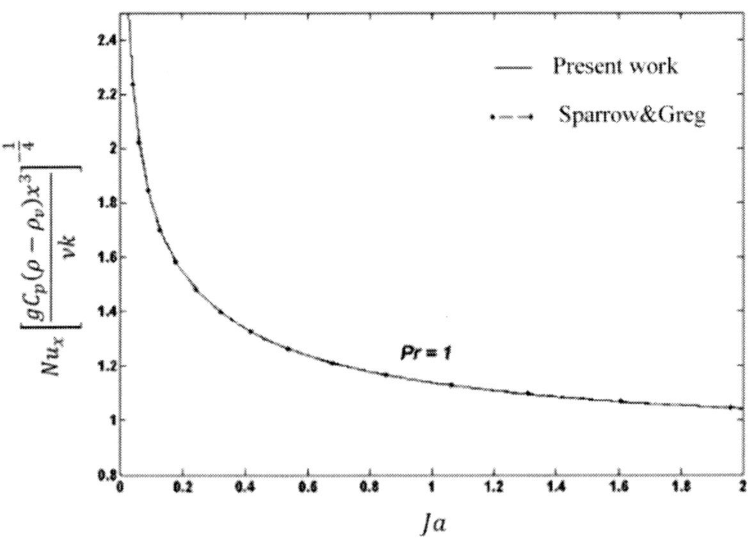

Figure 7. Local heat transfer versus *Ja* number for Pr=1.

So, in this section via the mentioned analysis we have tried to develop the previous works done by researchers and express some details of laminar film condensation phenomenon on a vertical plat.

4. Laminar Film Condensation on a Vertical Plate with Assisted Interfacial Shear Stress

As it was stated before, shear stress on the liquid-vapor interface is neglected in Nusselt theory. So, in the next step we try to restudy the problem of condensation on a flat plate with considering interfacial shear forces focusing on more details than the previous studies such as [66]. Two limiting cases will be introduced regarding the definition of boundary layer thickness as condensation on a flat plate with high gravitational forces compared to interfacial shear ones and condensation on a plate with high interfacial shear forces compared to gravitational ones. Also, a relation for the case of the two forces having comparable values and therefore not being in any of the two categories will be introduced and its validity will be discussed.

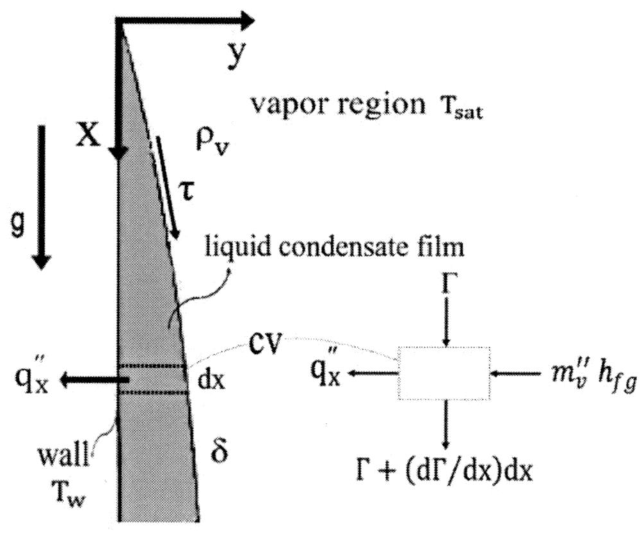

Figure 8. Physical model and coordinate system for condensate film flow on a vertical plate.

4.1. Physical Model

The physical model and coordinate system under considerations have been shown in figure 8. A saturated pure vapor with saturation temperature, T_{sat}, flows on a horizontal flat plate and imposes an associated interfacial shear, τ, which is supposed to be constant over the whole interfacial length. The wall temperature (T_w) is uniform and below the saturation temperature. Thus, condensation occurs on the wall and continues liquid film runs downward over the plate under the combined effects of gravity and the vapor interfacial shear force.

4.2. Mathematical Model

The assumptions that will be employed in the formulation of the problem are that the condensate film flow is laminar and steady state. Also, the inertia effect and viscous dissipation is ignored and compared with the conduction in the y-direction, conduction in the x-direction is negligible. Therefore, for a laminar, steady state condensate film with constant fluid properties, the boundary layer equations governed by the basis of conservation principles (mass, momentum, and energy) are:

$$\frac{\partial u}{\partial x}+\frac{\partial v}{\partial y}=0 \tag{39}$$

$$\mu\frac{\partial^2 u}{\partial y^2}=-g_x(\rho-\rho_v) \tag{40}$$

$$k\frac{\partial^2 T}{\partial y^2}=0 \tag{41}$$

Note that the momentum equation (equation (40)) reflects a balance of viscous and gravitational forces through its left and right terms, respectively.

It is further assumed that a shear force exists at the interface which was ignored in section 3. Thus, the boundary conditions subjected to the above equations are:

$$at \quad y=0\,, \quad u=0\,, \quad T=T_w \tag{42}$$

$$at \quad y = \delta\,, \quad \frac{\partial u}{\partial y} = \frac{\tau}{\mu}\,, \quad T = T_{sat} \tag{43}$$

It should be noted that that when τ =0, the problem is simplified to the classical theory of Nusselt [1] for condensation where no vapor shear is considered to exert upon the condensate.

By integrating the governing equations with respect to the mentioned boundary conditions, the dimensionless temperature and velocity profiles have been obtained as below:

$$\frac{T - T_w}{T_{sat} - T_w} = \frac{y}{\delta} \tag{44}$$

$$\frac{u}{V_{shear}} = Re_{\tau,\delta}\left[\frac{Ri_\delta}{Pr}\left(\frac{y}{\delta} - \frac{1}{2}\frac{y^2}{\delta^2}\right) + \frac{1}{2}\frac{y}{\delta}\right] \tag{45}$$

where

$$Re_{\tau,\delta} = \frac{V_{shear}\delta}{v}\ ; \quad Ri_\delta = \frac{Ar_\delta}{Re_{\tau,\delta}^2} \tag{46}$$

and

$$Ar_\delta = \frac{g(\rho - \rho_v)\delta^3 Pr}{\rho v^2}\ ; \quad V_{shear} = \left(\frac{2\tau}{\rho}\right)^{\frac{1}{2}} \tag{47}$$

In the above relations $Re_{\tau,\delta}$, V_{shear}, and Ri_δ are shear Reynolds number due to the shear effects of the vapor, shear velocity, and Richardson number. The shear velocity accounts for the assumed shear force at the liquid-vapor interface. Note here that the length scale in all dimensionless numbers presented above is the thickness of the liquid layer (δ).

What remains undefined in the above equations is the liquid film thickness (δ) which can be obtained by writing a balance of energy which was used in writing equation (7):

$$\dot{m}_v'' h_{fg} = k\frac{\Delta T}{\delta} \tag{48}$$

In the above relation, h_{fg} is the latent heat of vaporization and m_v'' is the condensate mass flow rate per unit area and can be calculated by applying the continuity equation to the differential element shown in figure 8:

$$\dot{m}_v'' = \frac{\partial \Gamma}{\partial x} \tag{49}$$

where Γ, is the condensate mass flow rate per unit width which is calculated by substituting the velocity profile from equation (45) in the following relation:

$$\Gamma(x) = \int_{y=0}^{y=\delta(x)} \rho u dy = \mu \left[\frac{Ar_\delta}{3Pr} + \frac{Re_{\tau,\delta}^2}{4}\right] \tag{50}$$

Hence, by differentiating the mass flow rate in the x direction m_v'' can be obtained:

$$\dot{m}_v'' = \rho V_{shear} \left[\frac{Ar_\delta}{PrRe_{\tau,\delta}} + \frac{Re_{\tau,\delta}}{2}\right] \frac{d\delta}{dx} \tag{51}$$

Substituting m_v'' in equation (48) will result in the following relation for the liquid film thickness (δ):

$$\delta^4 + \frac{2\rho V_{shear}^2}{3g(\rho - \rho_v)} \delta^3 = \frac{4k\nu\Delta T}{g(\rho - \rho_v)h_{fg}} x \tag{52}$$

Non-dimensionalizing equation (52) will result in:

$$\eta^3 \left(\eta + \frac{2}{3Ri_x}\right) = JPA_x \tag{53}$$

where

$$\eta = \frac{\delta}{x}; \quad JPA_x = \frac{4Ja}{PrAr_x} \tag{54}$$

and *Ja* was introduced before in equation (8).

The dimensionless numbers Ri_x, Ar_x and $Re_{\tau,x}$ have been introduced previously in non-dimensionalizing the velocity profile (equations (46) and (47));

however, the length scale here is x, not the thickness of the liquid layer (δ). Note that the Richardson number, Ri_x, can be counted as a ratio between the gravitational and interfacial shear forces. High values of Ri_x will represent the case of condensation on a vertical flat plate with negligible interfacial forces compared to the gravitational ones. Oppositely, low values of Ri_xwill account for the case of high interfacial shear forces compared to the gravitational ones. Considering the gravitational forces negligible, this case is similar to the problem of condensation on a horizontal flat plate where the liquid velocity is the result of interfacial shear forces.

4.3. Film Thickness

Equation (53) is the starting point for any analysis of forced convection condensation on an external surface. Since an analytical solution for η cannot be achieved from this equation, at this stage of the calculations, based on the definition of Ri_x, two limiting cases will be introduced.

4.3.1. Special Case I: $Ri_x \gg 1$

As it was mentioned above, high values of Ri_x represent the case of condensation on a flat plate with negligible interfacial forces compared to the gravitational forces. This case is similar to the Nusselt theory that no shear force is assumed to exist at the liquid vapor interface. In this case, equation (53) can be simplified to:

$$\eta^4 = JPA_x \tag{55}$$

This leads to an analytical solution for η:

$$\eta = (JPA_x)^{1/4} = \left(\frac{4Ja}{PrAr_x}\right)^{1/4} \tag{56}$$

4.3.2. Special Case II: $Ri_x \ll 1$

As it is expected, this case will account for the case of condensation on a flat plate with high interfacial shear forces compared to the gravitational ones. So, equation (53) can be simplified to:

$$\eta^3 \left(\frac{2}{3Ri_x}\right) = JPA_x \tag{57}$$

This leads to an analytical solution for η:

$$\eta = \left(\frac{3Ri_x\,JPA_x}{2}\right)^{1/3} = \left(\frac{6\,Ja}{\Pr Re_{\tau,x}^2}\right)^{1/3} \tag{58}$$

For the values of Ri_x which do not fit in the above categories, the situation is more complex and the solution for η cannot be achieved analytically. Therefore, in this case, equation (53) has been solved numerically. It should be noted that equation (53) is a fourth degree polynomial and has got four roots. In the numerical solution for η there is only one real positive root which has been considered valid for the problem. Therefore, continuing the calculations from equation (53) for these values of Ri_x, it can be obtained that:

$$\eta = f(Ri_x, JPA_x) \tag{59}$$

Assuming the *f* function as:

$$f(x, y) = Cx^m y^n$$

The dimensionless liquid film thickness can be gained as:

$$\eta = C(1.5Ri_x)^m (JPA_x)^n \tag{60}$$

For the two limiting cases, the values of *m* and *n* have already been calculated analytically (equations (56) and (58)):

$$case\ I\ (Ri_x \gg 1):\ \ C = 1,\ \ m = 0,\ \ n = 0.25 \tag{61}$$

$$case\ II\ (Ri_x \ll 1):\ \ C = 1.14,\ \ m = 0.33,\ \ n = 0.33 \tag{62}$$

For the values of Ri_x between the two limiting cases, that are when the two forces have comparable values to each other, a relation has been obtained from the numerical values:

$$\eta = 1.36 Ri_x^{0.2} JPA_x^{0.3} \tag{63}$$

which shows acceptable agreement with the two limiting cases. As it is expected, the powers of the two dimensionless groups (*m* and *n*) have got values between the two limiting cases.

4.4. Flow Rate

The dimensionless mass flow rate in the condensation phenomenon may be expressed in terms of a Reynolds number and the Reynolds number can be defined as:

$$Re_\delta = \frac{4\Gamma}{\mu} \tag{64}$$

Putting the liquid mass flow rate, Γ, presented in equation (50), the *Re* number may be expressed as below:

$$Re_\delta = \frac{4}{3}\frac{Ar_\delta}{Pr} + Re_{\tau,\delta}^2 \tag{65}$$

As it is seen, the mass flow rate, Γ, is affected by both gravitational effects of the fluid and the vapor shear velocity. In other words, unlike the case of Nusselt theory where the only important factor in the flow regime was the gravitational effect, in the case of considering the interfacial shear forces, another factor due to these forces should be taken into account ($Re_{x,\delta}$). As the shear velocity was previously assumed to represent the interfacial effects, here the shear Reynolds number ($Re_{x,\delta}$) is the representative factor for it. It can be noticed that, for high values of Ar_δ compared to $Re_{x,\delta}$, Re_δ can be obtained as:

$$Re_{\tau,\delta} \ll \frac{Ar_\delta}{Pr}, \quad Re_\delta = \frac{4}{3}\frac{Ar_\delta}{Pr} \tag{66}$$

which means the only effective factor on the fluid regime is the gravitational force. This is expected as it is the result gained by the Nusselt theory. But it can be expected, for the opposite case of having high interfacial shear force compared to the gravitational one, the gravitational force does not have a big role on the mass flow rate as the interfacial one does. In this case Re_δ can be obtained as

$$Re_{\tau,\delta} \gg \frac{Ar_\delta}{Pr}, \quad Re_\delta = Re_{\tau,\delta}^2 \tag{67}$$

This case is similar to the case of condensation on a horizontal flat plate with forced vapor flow.

4.5. Nusselt Number

As in Nusselt theory, interpreting a local heat transfer coefficient gives:

$$h_x = \frac{k}{\delta} = \frac{k}{\eta x} \tag{68}$$

Therefore,

$$Nu_x = \frac{1}{\eta} \tag{69}$$

The average Nusselt number can also be obtained as:

$$\overline{Nu} = \frac{\overline{h_L} L}{k} \tag{70}$$

where average heat transfer coefficient can be defines as:

$$\overline{h_L} = \frac{1}{L}\int_0^L h_x dx \tag{71}$$

4.5.1. Special Case I: Ri_x>>1

Local heat transfer coefficient can be gained as:

$$h_x = \frac{k}{x}\left[\frac{\Pr Ar_x}{4\,Ja}\right]^{1/4}$$

Therefore,

$$Nu_x = \left[\frac{\Pr Ar_x}{4Ja}\right]^{1/4} \; ; \; \overline{Nu} = \frac{\overline{h_L}L}{k} = \frac{4}{3}\left[\frac{\Pr Ar_L}{4Ja}\right]^{1/4} \tag{72}$$

This result is similar to the result of Nusselt theory [1].

4.5.2. Special Case II: Ri_x<<1

For this case, local heat transfer coefficient can be introduces as:

$$h_x = \frac{k}{x}\left[\frac{\Pr Re_{\tau,\delta}^2}{6Ja}\right]^{1/3}$$

and local Nusselt number can be gained as follow:

$$Nu_x = \left[\frac{\Pr Re_{\tau,x}^2}{4Ja}\right]^{1/3} \; ; \; \overline{Nu} = \frac{\overline{h_L}L}{k} = \frac{3}{2}\left[\frac{\Pr Re_{\tau,L}^2}{6Ja}\right]^{1/3} \tag{73}$$

Due to the relation obtained for the values of Ri_x between the two limiting cases (equation (63)), for these values of Ri_x, it can be said that:

$$h_x = 0.74\frac{k}{x}Ri_x^{-0.2}JPA_x^{-0.3}$$

Therefore,

$$Nu_x = 0.74Ri_x^{-0.2}JPA_x^{-0.3} \; ; \; \overline{Nu} = 1.05Ri_L^{-0.2}JPA_L^{-0.3} \tag{74}$$

4.6. Results and Discussion

The effect of Ri_x on the velocity profile has been illustrated in figure 9. As it is seen, for constant values of Archimedes number, the increase in Re_δ and consequently the decrease in Ri_δ will lead to an increase in the value of $u\delta/\nu$. It will be shown later that high interfacial forces (low Richardson number) will result in thinning the condensate layer (δ). Assuming constant properties for the liquid (constant ν), it has been resulted that the velocity profile has greater values in the cases with higher interfacial forces compared to the Nusselt case were no shear stress is assumed to exert upon the condensate layer.

The effect of *Ja/Pr* on the dimensionless film thickness is shown in figure 10. The diagram shows the analytical values of the two limiting cases, considering only one of the two gravitational and interfacial shear forces (Ri_x>>*1* and Ri_x <<*1*, respectively) as well as the numerical values (two cases of Ri_x=*10* and Ri_x =*0.1*). A significant characteristic of this diagram is the lower slope of the diagrams related to higher values of Ri_x compared to the ones for lower values of Ri_x. As was mentioned before, the values of dimensionless liquid film thickness vary with $(Ja/Pr)^n$ (equations (56) and (58)) where the power *n* being 1/4 for high values of Ri_x and 1/3 for lower ones, explains the differences in the slopes of the diagrams.

As it is expected, higher values of *Ja/Pr* will lead to higher values of the dimensionless liquid film thickness. Also, it has been observed that the dimensionless liquid film thickness will decrease with decreasing the value of Ri_x. In other words, for constant values of the gravitational forces, increasing the interfacial shear forces will decrease the rate of liquid film thickness growth. Also, a comparison has been made for two values of shear Reynolds number in the case of higher interfacial shear forces (Re_x=14 and 20). The results show that in the case of having high interfacial shear force, the higher the value of shear Reynolds number is, the lower the liquid film growth is. The comparison has not been made for the case of low interfacial forces as in that case the liquid film thickness is not a function of shear Reynolds number.

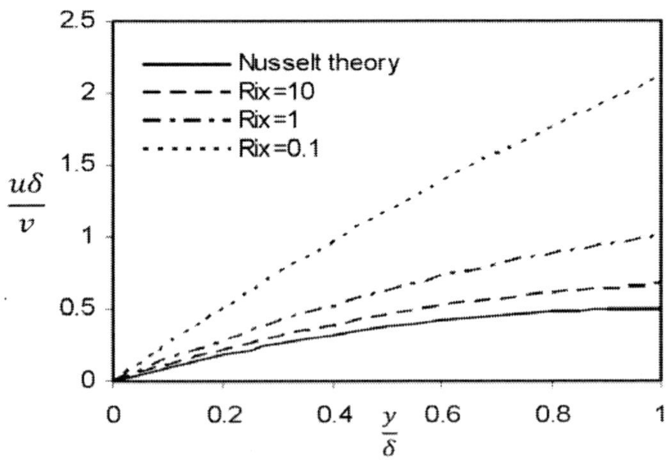

Figure 9. The effect of Ri_x number on the velocity profile for Ar_x=*1*.

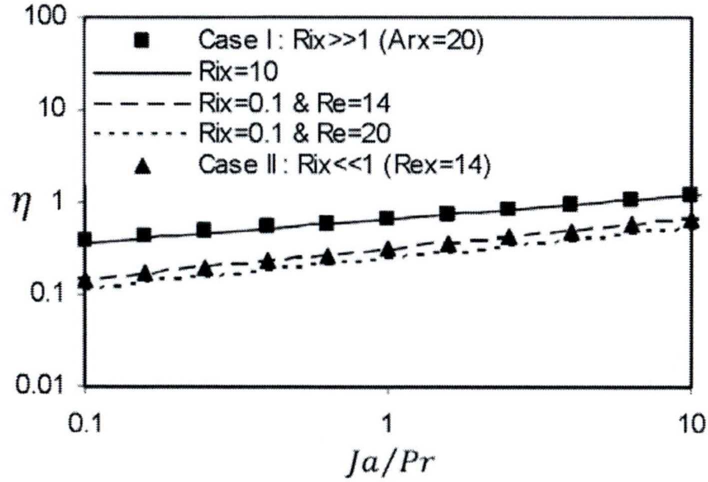

Figure 10. The effect of *Ja/Pr* on the dimensionless film thickness (η).

Comparing the analytical results to the numerical ones, it has been seen that slight differences of less than 3% for high values of *Ja/Pr* and 1% for low values of *Ja/Pr* exist between the two cases of Ri_x<<*1* (analytical) and Ri_x=*0.1* (numerical). Similar differences are recognized for the limiting case of Ri_x>>*1* and the numerical values for Ri_x=10. In other words, for the values of Ri_x being higher than 10 or lower than 0.1, the analytical solutions for the first (Ri_x>>*1*) and the second (Ri_x<<*1*) limiting cases may be applied respectively with a good approximation degree instead of the numerical calculations.

The effect of Ri_x on the dimensionless liquid film thickness is shown in figure 11. The first noticeable characteristic of the diagram is the increase in the value of the dimensionless liquid film thickness for high values of Ri_x, representing the results of Nusselt theory. So, it has been concluded that for an extremely low Richardson number, in other words when a vapor with downward direction is forced past a cool wall, it will drag the liquid film along and will make it thinner. Also, it has been observed that for various values of JPA_x, same values for $\eta/[4\,Ja/\Pr Ar]^n$ have been resulted. In other words, the diagrams are only functions of Ri_x. As it can be seen, for very low values of Ri_x (lower than about 0.2) the analytical values of the second limiting case (Ri_x<<*1*) are very close to the numerical values (difference of less than 9%). They have almost the same values for Ri_x<*0.1*. This means that for Ri_x<*0.1*, the analytical limiting solution for the second case (Ri_x <<*1*) can be applied with a good degree of approximation (error of less than 5%). Similarly, for values of Ri_x higher than

Table 1. Error analysis of the analytical relations derived for various ranges of Ri_x

Range	The Liquid Film Thickness	Maximum Error
$Ri_x < 0.1$	$\eta = 1.14(Ri_x)^{0.33}(JPA_x)^{0.33}$	5%
$0.1<Ri_x<0.2$	$\eta = 1.14(Ri_x)^{0.33}(JPA_x)^{0.33}$	9%
$0.2<Ri_x<1.6$	$\eta = 1.36(Ri_x)^{0.2}(JPA_x)^{0.3}$	3%
$1.6<Ri_x<10$	$\eta = (JPA_x)^{0.25}$	15%
$10<Ri_x$	$\eta = (JPA_x)^{0.25}$	3%

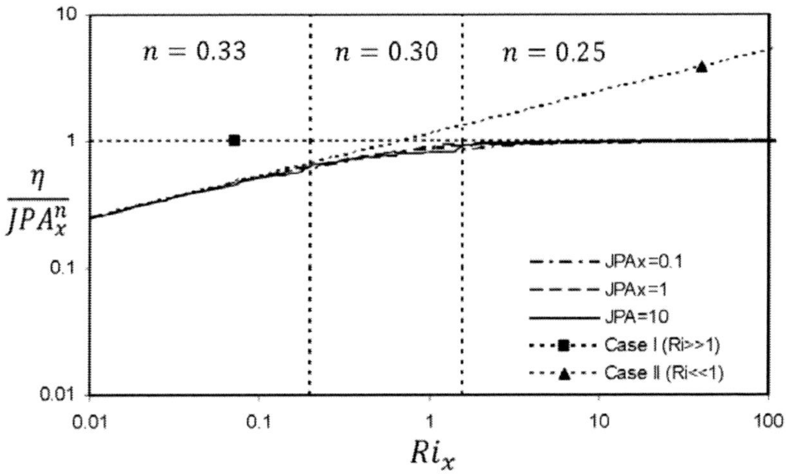

Figure 11. The effect of Ri_x on the dimensionless liquid film thickness.

about 1.6, the analytical values of the first limiting case ($Ri_x >> 1$) show differences of about 15% to the numerical values and almost coincide the numerical values for $Ri_x > 10$. This means that for $Ri_x > 10$, the analytical limiting solution for the second case ($Ri_x >> 1$) can be applied with an error of less than 3%. For values of $0.2\leq Ri_x \leq 1.6$, the dimensionless liquid film thickness can be obtained from the diagram due to the value of n=0.3 gained from the numerical results. The recommended relationships for liquid film thickness and their relative errors presented above have been summarized in table 1.

Figure 12 shows the validity of the relationship gained numerically for the liquid film thickness in the range of $0.2\leq Ri_x \leq 1.6$ (equation (63)) in comparison with the two limiting cases ($Ri_x<<1$ and $Ri_x>>1$). Here, the numerical values for the dimensionless liquid film thickness (η) of the previous figure are non-dimensionalized by $C(Ri_x)^m$. Note that the values of C, m and n are the ones prescribed previously for each criterion. Therefore, the more accurate the liquid film thickness relationships are for each criterion, the closer the dimensionless values of the diagrams get to one. It has been shown that the numerical values for very low and very high values of Ri_x show good agreement with the analytical values (equations (56) and (58)). As for the values of Ri_x between the two limiting cases, equation (63) can be counted as an approximate relation for calculating the dimensionless liquid film thickness with the maximum error of 15% near Ri_x=*1.6* (discussed previously).

Figure 13 depicts the comparison made between the values of Nusselt number from the results in this chapter (Nu_x) and the ones from the Nusselt theory ($Nu_{x,N}$) for a wide range of Ri_x and at different shear Reynolds numbers. It has been observed that for high values of Ri_x representing low values of interfacial shear forces compared to the gravitational ones, the numerical results have similar values to the results of Nusselt theory in which no shear stress is assumed to exert upon the condensate. As it was anticipated, the numerical values for low Ri_x are higher than the values of the Nusselt theory since high values of interfacial shear forces in this region result in a thinner liquid film thickness and the enhancement of the heat transfer coefficient.

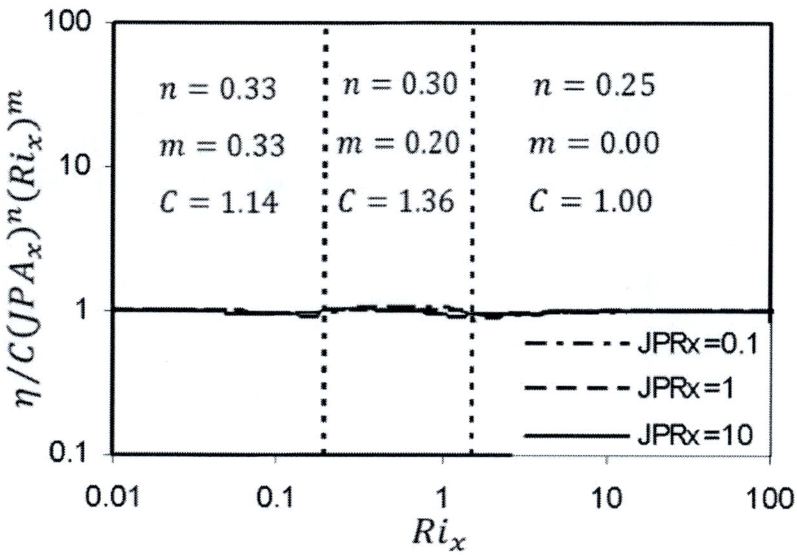

Figure 12. The validity of the η relation for 0.2≤ Ri_x ≤1.6 in comparison with the two limiting cases (Ri_x<<*1* and Ri_x>>*1*).

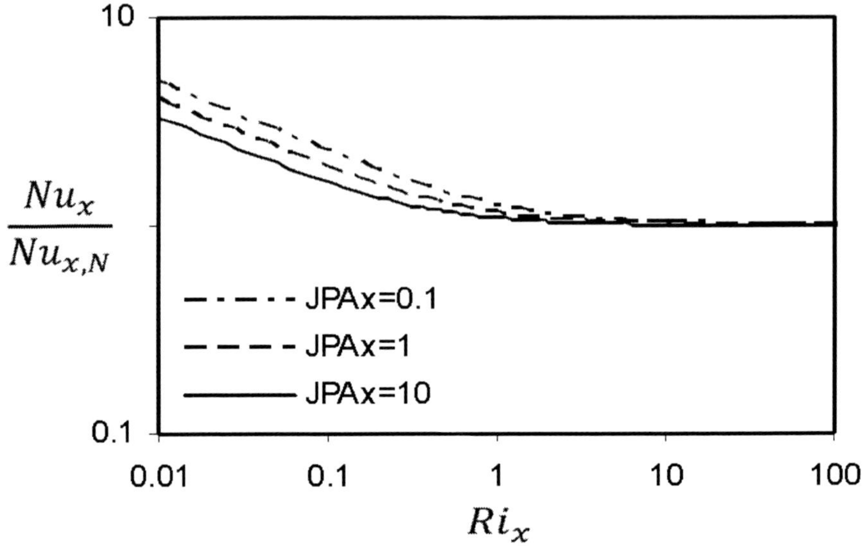

Figure 13. The effect of Ri_xon the Nu_x number.

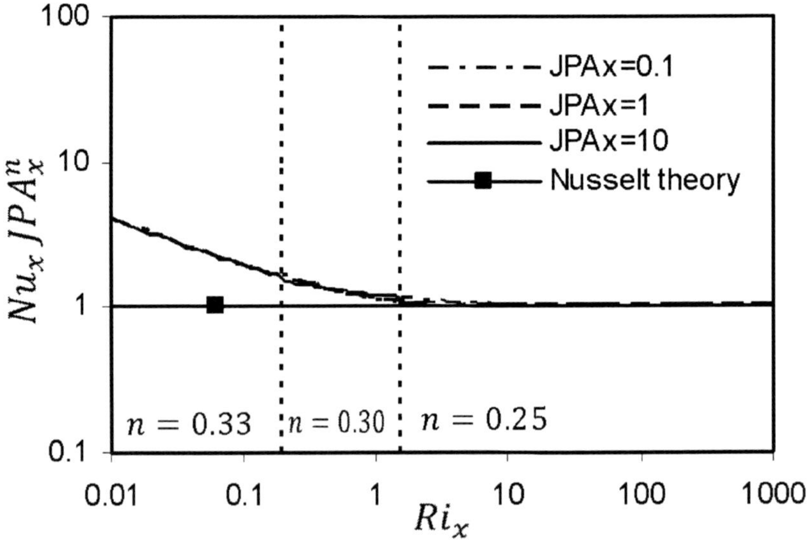

Figure 14. The effect of Ri_x on the Nu_x number.

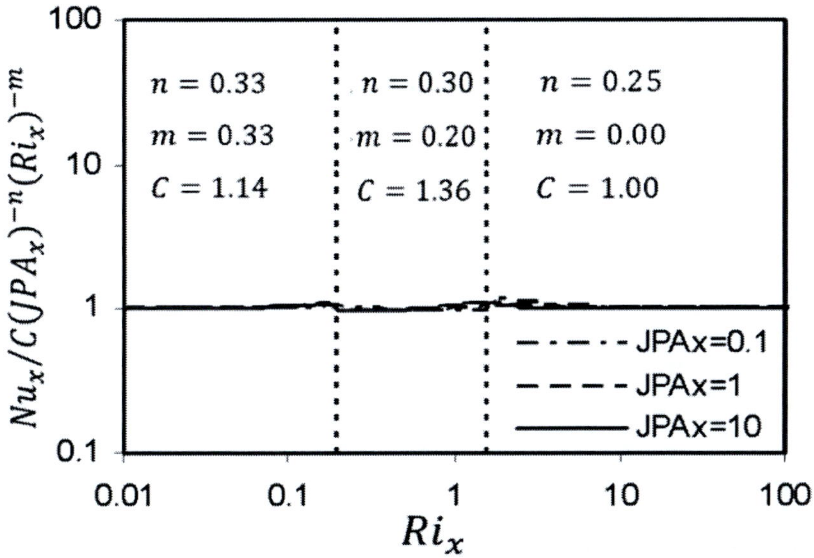

Figure 15. The validity of the Nu_x relation for 0.2≤ Ri_x ≤1.6 in comparison with the two limiting cases(Ri_x<<*1* and Ri_x>>*1*).

Figure 14 illustrates the effect of Ri_x on *Nu* number. Similar to figure 11, for various values of JPA_x, same values for $(Nu_x JPA_x^n)$ have been resulted. In other words, the diagrams are only the functions of Ri_x. It has been observed that for high values of Ri_x representing low values of interfacial shear forces compared to the gravitational ones, the numerical results overlap the results of Nusselt theory. The numerical values for low values of Ri_x are higher than the Nusselt's results since in this case a high shear stress exists at the interface which has not been considered in the Nusselt theory. Also, thinner liquid film resulted by the high interfacial shear force assumed in the second limiting (Ri_x<<*1*) case will enhance the heat transfer coefficient.

The validity of the Nu_x relation for 0.2≤ Ri_x ≤1.6 (equation (65)) in comparison with the two limiting cases (Ri_x<<*1* and Ri_x>>*1*) has been displayed in figure 15. As it has been shown, the numerical values of Nu_x are divided by the values of Nu_x calculated analytically for each criterion (equations (72), (74) and (65)). In this diagram a maximum error of 15% exists near Ri_x=1.6. However, for very low and very high values of Ri_x (Ri_x<<*0.1* and Ri_x>>*10*, respectively) the numerical values show a good agreement with the analytical values calculated from equations (72) and (74) (as mentioned before, error of less than 5% and 3%, respectively).

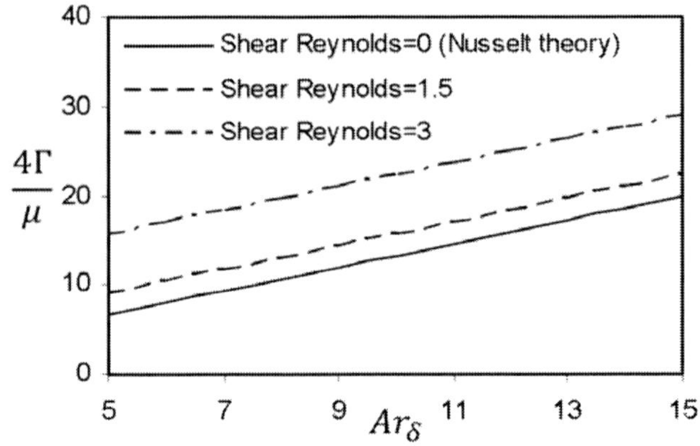

Figure 16. The variation of liquid mass flow rate per unit width (Γ) with Archimedes number (Ar_δ) when *Pr=1.*

Figure 16 indicates the variation of liquid mass flow rate per unit width (Γ) with Archimedes number (Ar_δ). It has been seen that for constant values of dynamic viscosity (μ) and Archimedes, increasing the interfacial shear forces will result in an increasing shift (equal to $Re_{\tau,\delta}^2$) in the liquid mass flow rate per unit width. It has been noticed that although the interfacial shear forces have a decreasing effect on the liquid film thickness, the increase in the velocity values will have an overall increasing effect on the liquid mass flow rate.

Thus, in this section we tried to alter Nusselt theory by considering the shear stress at the liquid-vapor interface and analyze the problem for two selected limiting cases as mentioned above. So, a modification on simple Nusselt theory for laminar film condensation on a vertical plate has been done.

5. Laminar Film Condensation on a Vertical Plate with Assisted Vapor Flow

Now we will try to develop the problem of condensation on a flat plate with vapor free stream velocity focusing on more details than the previous studies [66]. A general relation in the form of an ordinary differential equation will be derived for calculating the liquid film thickness. The calculations for the liquid film thickness will be followed by the Nusselt calculations and will be discussed for various values of Reynolds number.

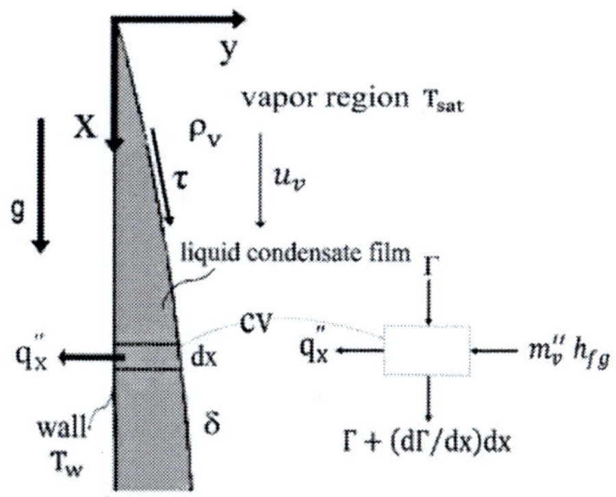

Figure 17. Physical model and coordinate system for condensate film flow on a vertical plate.

5.1. Physical Model

The physical model and coordinate system are similar to what was presented for the problem of laminar film condensation on a vertical plate with assisted shear stress showed in figure 8 except that for the condensation with vapor flow, a vapor velocity exist on the interface of liquid-vapor. This new condition has been displayed in the figure 17.

5.2. Mathematical Model

Mathematical model and the assumptions employed in the formulation of the problem are the same as which were introduced for the problem of film condensation on a flat plate with assisted interfacial shear stress in section 4. The only difference is that a vapor boundary layer is assumed to exist on the condensate layer. Therefore, as it was mentioned before for a laminar, steady state condensate film with constant fluid properties, the boundary layer equations governed by the basis of conservation principles are the same as equations (39), (40), and (41) and boundary related conditions are similar to equations (42) and (43).

The shear stress τ in the velocity boundary condition at the liquid vapor interface can be replaced by its definition due to the characteristics of the vapor flow as below:

$$\tau = \frac{1}{2} C_f \rho_v (u_v - u_i) \tag{75}$$

where C_f, u_i and u_v are the coefficient of friction, interfacial and potential flow velocities, respectively. The coefficient of friction, C_f, can be gained as:

$$C_f = 0.664 Re_{v,x}^{-\frac{1}{2}} \tag{76}$$

and

$$Re_{v,x} = \frac{(u_v - u_i)x}{\upsilon_v} \tag{77}$$

where $Re_{v,x}$ is the vapor Reynolds number. Note that when $u_v = u_i$, the problem is simplified to the classical theory of Nusselt for condensation [1] where no vapor shear is considered to exert upon the condensate layer.

For the prescribed governing equations and boundary conditions, the dimensionless temperature and velocity profiles have been obtained as below:

$$\frac{T - T_w}{T_{sat} - T_w} = \frac{y}{\delta} \tag{78}$$

$$\frac{u}{u_v - u_i} = \frac{Ar_L}{\Pr Re_L} \bar{\delta}^2 \left(\frac{y}{\delta} - \frac{1}{2} \frac{y^2}{\delta^2} \right) + 0.332 Re_{v,L}^{\frac{1}{2}} \bar{\delta} \bar{x}^{-\frac{1}{2}} \left(\frac{\mu_v}{\mu} \right) \left(\frac{y}{\delta} \right) \tag{79}$$

where

$$\bar{x} = \frac{x}{L}; \quad \bar{\delta} = \frac{\delta}{L}; \quad Re_L = \frac{(u_v - u_i)L}{\nu} \tag{80}$$

In the above relations, L and Re_L are the plate length and relative Reynolds number due to the velocity difference of the vapor. Archimedes number was mentioned before in equation (47). It should be noted that Re_L presented above is the liquid Reynolds number which can be converted to vapor Reynolds number ($Re_{v,x}$) by substituting ν by ν_v.

Equation (79) can be written in the form of:

$$\frac{uL}{v} = \frac{Ar_L}{Pr}\bar{\delta}^2\left(\frac{y}{\delta} - \frac{1}{2}\frac{y^2}{\delta^2}\right) + 0.332Re_{v,L}^{\frac{3}{2}}\bar{\delta}\bar{x}^{-\frac{1}{2}}\left(\frac{\mu_v}{\mu}\right)\left(\frac{v_v}{v}\right)\left(\frac{y}{\delta}\right) \quad (81)$$

The first term in the right hand side of equation (81) accounts for the gravitational forces. As can be noticed, assuming an assisted vapor flow for the problem adds a second term to the velocity profile. Therefore, higher values of the velocity profile are expected for the case of having a vapor flow and the higher the value of vapor free stream velocity (higher $Re_{v,L}$) is, the higher the values of the velocity profile across the liquid film thickness are.

Also, the term corresponding to the gravitational forces is of second degree of y/δ and the term corresponding to the vapor velocity effects is linear. Consequently, for the limiting case of high Ar_L number compared to $Re_{v,L}$, the effect of the vapor free stream velocity (the second term) becomes very small and the velocity profile versus y/δ tends to take a polynomial form while for the inverse case of low Ar_L number compared to $Re_{v,L}$, it will take a linear form.

Furthermore, the relative Reynolds number can be expressed in the form of:

$$Re_L = Re_L'\left(1 - \frac{u_i}{u_v}\right) \quad (82)$$

where

$$Re_L' = \frac{u_v L}{v} \quad (83)$$

It has been seen that for very low values of interfacial velocity (u_i) compared to vapor free stream velocity (u_∞), the term $\frac{u_i}{u_v}$ in equation (82) can be negligible. Therefore:

$$Re_L \approx Re_L' \quad (84)$$

5.3. Film Thickness

What remains undefined in the above equations is the liquid film thickness (δ) and as it was stated before, it can be obtained by writing a balance of energy which was used in equations (7) and (48).

The condensate mass flow rate per unit width (Γ) which was introduced in equation (50) can also be gained here:

$$\Gamma(x) = \int_{y=0}^{y=\delta(x)} \rho u dy = \mu \frac{Ar_L}{3Pr} \bar{\delta}^3 + 0.166 Re_{v,L}^{\frac{3}{2}} \bar{\delta}^2 \bar{x}^{-\frac{1}{2}} \left(\frac{\nu_v}{\nu}\right) \mu_v \tag{85}$$

Hence, by differentiating the mass flow rate per unit width in the x direction, the condensate mass flow rate per unit area ($\dot{m}_v''$) can be obtained as:

$$\dot{m}_v'' = \frac{\mu}{L} \left\{ \left[\frac{Ar_L}{Pr} \bar{\delta}^2 + 0.332 Re_{v,L}^{\frac{3}{2}} \bar{x}^{-\frac{1}{2}} \bar{\delta} \left(\frac{\nu_v}{\nu}\right) \left(\frac{\mu_v}{\mu}\right) \right] \frac{d\bar{\delta}}{d\bar{x}} - 0.083 Re_{v,L}^{\frac{3}{2}} \bar{x}^{-\frac{3}{2}} \bar{\delta}^2 \left(\frac{\nu_v}{\nu}\right) \left(\frac{\mu_v}{\mu}\right) \right\} \tag{86}$$

So by substituting $\dot{m}_v''$ in the balance of the energy, following relation for the liquid film thickness (δ) has been gained:

$$\frac{d\bar{\delta}}{d\bar{x}} = \frac{D + B(\bar{x}, \bar{\delta})}{A(\bar{\delta}) + C(\bar{x}, \bar{\delta})} \tag{87}$$

where

$$A(\bar{\delta}) = \frac{Ar_L}{Pr} \bar{\delta}^3$$
$$B(\bar{x}, \bar{\delta}) = 0.083 Re_{v,L}^{\frac{3}{2}} \bar{x}^{-\frac{3}{2}} \bar{\delta}^3 \left(\frac{\nu_v}{\nu}\right) \left(\frac{\mu_v}{\mu}\right)$$
$$C(\bar{x}, \bar{\delta}) = 0.332 Re_{v,L}^{\frac{3}{2}} \bar{x}^{-\frac{1}{2}} \bar{\delta}^2 \left(\frac{\nu_v}{\nu}\right) \left(\frac{\mu_v}{\mu}\right)$$
$$D = Ja \times Pr \tag{88}$$

The dimensionless liquid film thickness (δ) can be gained numerically via the forth order Runge-Kutta method from equation (87).

5.4. Flow Rate

The dimensionless mass flow rate (local Reynolds number) has been defined before in equation (64). By putting Γ which was presented in equation (85), the Re_δ will be gained as:

$$Re_\delta = \underbrace{\frac{4Ar_L}{3Pr}\bar{\delta}^3}_{\text{Gravitational effect of the fluid}} + \underbrace{0.664Re_{v,L}^{\frac{3}{2}}\bar{\delta}^2\bar{x}^{-\frac{1}{2}}\left(\frac{\nu_v}{\nu}\right)\left(\frac{\mu_v}{\mu}\right)}_{\text{Effect of vapor free stream velocity}} \tag{89}$$

Like section 4.4, it is seen that Γ is affected by both gravitational effects of the fluid and the vapor free stream velocity. So, unlike the case of Nusselt theory, another factor due to these forces should be taken into account ($Re_{v,L}$). It can be noticed that for high values of Ar_L compared to $Re_{v,L}$, Re_δ can be obtained as:

$$Re_\delta = \frac{4Ar_L}{3Pr}\left(\frac{\delta}{L}\right)^3 \tag{90}$$

For the opposite case of having vapor flow with a high velocity, Re_δ can be defined as:

$$Re_\delta = 0.664Re_{v,L}^{\frac{3}{2}}\bar{\delta}^2\bar{x}^{-\frac{1}{2}}\left(\frac{\nu_v}{\nu}\right)\left(\frac{\mu_v}{\mu}\right) \tag{91}$$

This case is similar to the case of condensation on a horizontal flat plate with forced vapor flow.

5.5. Nusselt Number

As it was said in section 4.5, in Nusselt theory interpreting a local heat transfer coefficient gives:

$$h_x = \frac{k}{\delta} = \frac{k}{\bar{\delta}L} \tag{92}$$

Therefore

$$Nu_L = \frac{1}{\bar{\delta}} \tag{93}$$

By having the numerical values of equation (87), Nu_L can be obtained. The average Nusselt number can also be achieved as:

$$\overline{Nu} = \frac{\overline{h_L} L}{k} = \int_0^1 \frac{1}{\bar{\delta}} d\bar{x} \tag{94}$$

and $\overline{h_L}$ is the same as which was introduced in equation (71).

5.6. Results and Discussion

The numerical solution of equation (87) for calculating the liquid film thickness is carried out for $Ar_L/Pr = 10^8$, $Re_L = 10^6$, $Re_{v,L} = 10^4$, $Ja = 0.1$ and $(v_v/v)(\mu_v/\mu) = 3.46$ except for the cases where different values are introduced.

The effect of $Re_{v,L}$ on the ratio of the velocity at the liquid vapor interface to the vapor free stream velocity has been shown in figure 18. It can be observed that the values of the interfacial velocity are very low compared to the vapor free stream velocity (about 1%). Therefore, approximating this velocity ratio (u_i/u_∞) to zero will simplify the numerical calculations with not adding a big deal of error (less than 1%).

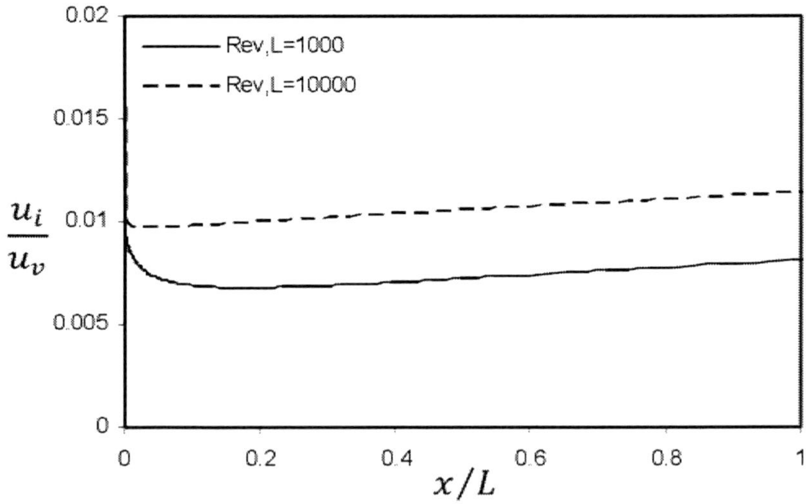

Figure 18. The effect of $Re_{v,L}$ on the ratio of the velocity at the liquid vapor interface to the vapor free stream velocity.

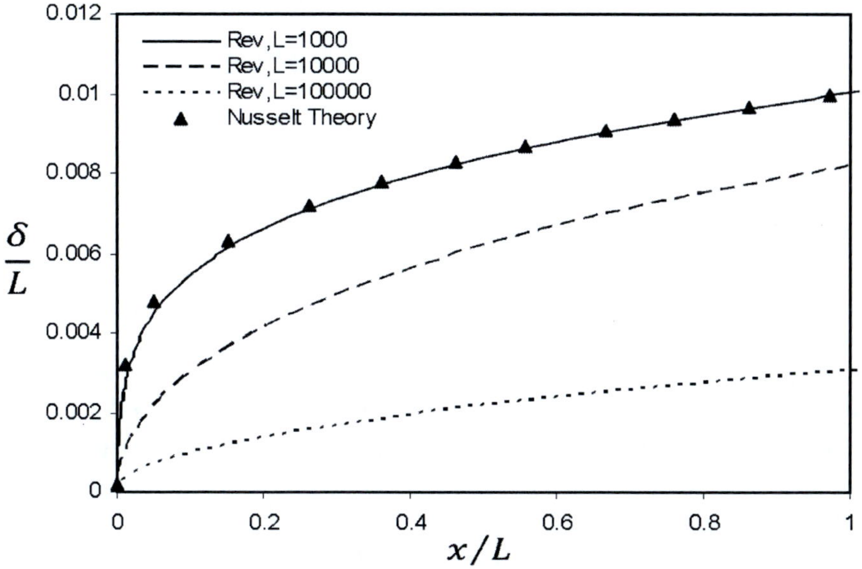

Figure 19. The effect of $Re_{v,L}$ on the dimensionless liquid film thickness (δ/L).

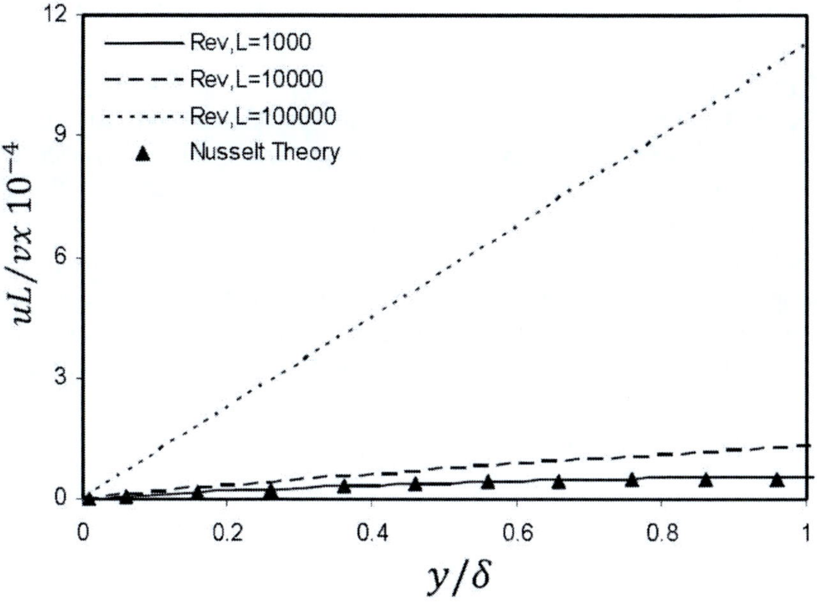

Figure 20. The effect of $Re_{v,L}$ on the velocity profile for $x/L = 1$.

The effect of $Re_{v,L}$ on the dimensionless liquid film thickness has been illustrated in figure 19. As it was expected, marching along the plate will have an increscent in the dimensionless liquid film thickness. Another characteristic of the diagram is the increase in the value of the dimensionless liquid film thickness for low values of $Re_{v,L}$ (lower than 1000). So, it has been concluded that when a vapor with downward direction is forced past a cool wall (high Reynolds number), it will drag the liquid film along and will make it thinner. Also, it has been observed that for low values of $Re_{v,L}(Re_{v,L} = 10^3$, when $Ar_L/Pr = 10^8$) representing low values of vapor free stream velocity, the numerical results overlap the results of Nusselt theory that no shear stress is assumed to exert upon the condensate layer. Furthermore, it has been seen that for all vapor Reynolds numbers, there is a sudden growth of the condensate layer (89°) at the beginning of the plate $(x/L \to 0)$ which may be reliable due to neglecting the advection terms in the momentum equation.

Figure 20 indicates the effect of $Re_{v,L}$ on the velocity profile. It has been seen that for constant values of $Ar_L/Pr(Ar_L/Pr = 10^8)$ the increase in $Re_{v,L}$ will lead to an increase in the value of uL/ν. As it was shown in figure 19, high vapor free stream velocities will result in thinning the condensate layer. Therefore, assuming constant properties for the liquid (constant ν), figure 20 shows that the velocity profile has higher values in cases with higher vapor free stream velocity (higher $Re_{v,L}$) compared to the Nusselt case. This is expected as the thinner condensate layer will have an enhanced heat transfer coefficient and consequently an enhanced condensate mass flow rate. Due to the relation for the condensate mass flow rate and having a thinner condensate layer, the increased mass flow rate for high vapor free stream velocity is possible only when the velocity profile has higher value which has been illustrated in figure 20. Furthermore, the velocity profile is the sum of two terms corresponding to gravitational and vapor free stream velocity effects (equation (81)). Due to the previous discussions made on equation (81) and as it has been shown in the figure, the velocity profile tends to take a polynomial form for the limiting case of low $Re_{v,L}$ number ($Re_{v,L} = 10^3$) compared to Ar_L, while it will take a linear form for the inverse case of high $Re_{v,L}$ number ($Re_{v,L} = 10^5$) compared to Ar_L.

The growth of the velocity profile along the plate length has been shown in figure 21. It has been observed that marching through x direction will have an increase in the values of the velocity profile. This was expected due to the increase in mass flow rate along the plate.

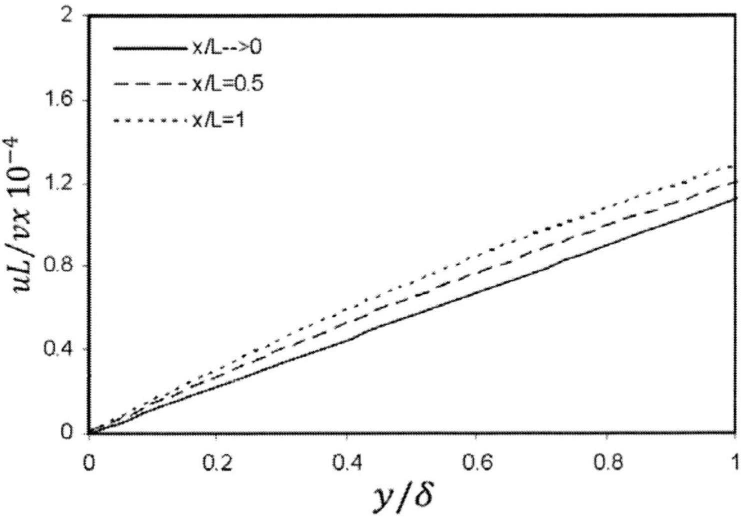

Figure 21. The growth of the velocity profile along the plate length when $Re_{v,L} = 10^4$.

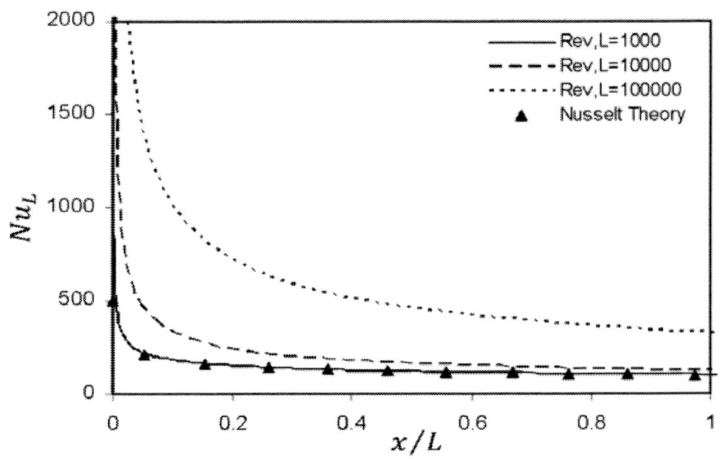

Figure 22. The effect of $Re_{v,L}$ on Nu number.

Figure 22 displays the effect of $Re_{v,L}$ on Nu_x number. It has been concluded that for low values of $Re_{v,L}$ ($Re_{v,L}$=1000) representing low values of vapor free stream velocity, the numerical results overlap the results of Nusselt theory. It was anticipated from the discussion made on figure 19; thinner liquid film thickness corresponding to higher vapor free stream velocity will enhance the heat transfer coefficient.

The effect of *Ja* on *Nu* number has been illustrated in figure 23. It has been shown that higher values of *Ja* number will result lower values of *Nu* numbers. As was mentioned previously, the values of dimensionless liquid film thickness vary with *Ja* (equations (87) and (88)). Similar to the Nusselt's calculations for the liquid film thickness, where the liquid film thickness varies with $Ja^{1/4}$, here as well, higher values of *Ja* will lead to higher values of dimensionless liquid film thickness and consequently a lower heat transfer coefficient (decrease from $(Nu_L)_{Ja=0.1} = 127$ to $(Nu_L)_{Ja=0.4} = 84$ at $\bar{x} = 1$).

Figure 24 depicts the variation of *Nu* number with Ar_L. For constant values of $Re_{v,L}$, increasing Ar_L will cause increscent in the *Nu* number. This result was anticipated as it was proved that higher values of Ar_L corresponding to high gravitational forces drag the liquid film down and making it thinner. Consequently, the thinner liquid film results the higher heat transfer coefficient (increase from $(Nu_L)_{Ar/Pr=1e6} = 103$ to $(Nu_L)_{Ar/Pr=1e10} = 313$ at $\bar{x} = 1$).

Figure 25 shows the variation of Γ with Reynolds number ($Re_{v,L}$). It has been observed that for a constant value of Archimedes number, increasing the vapor free stream velocity ($Re_{v,L}$) will result in an increase in the dimensionless liquid mass flow rate per unit width. Although the vapor free stream velocity have a descending effect on the liquid film thickness, the increase in the condensate velocity values will have an overall increasing effect on the liquid mass flow rate.

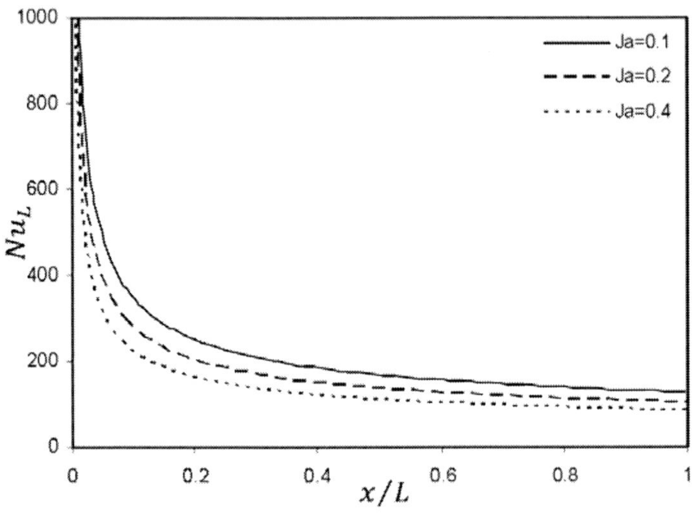

Figure 23. The effect of *Ja* on N*u* number.

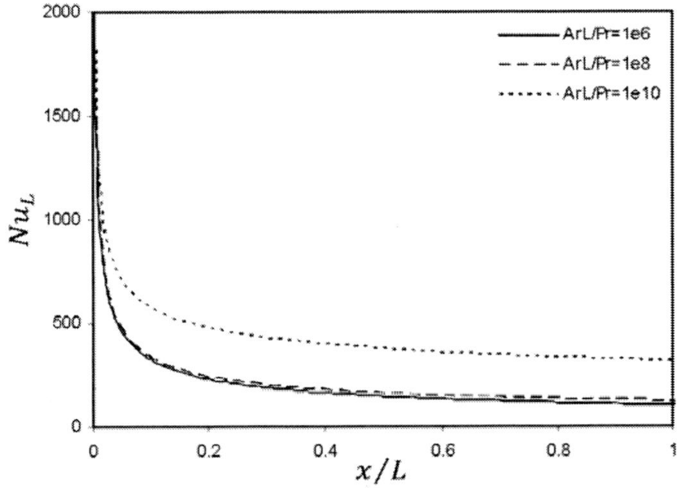

Figure 24. Variation of Nu number with Ar_L number.

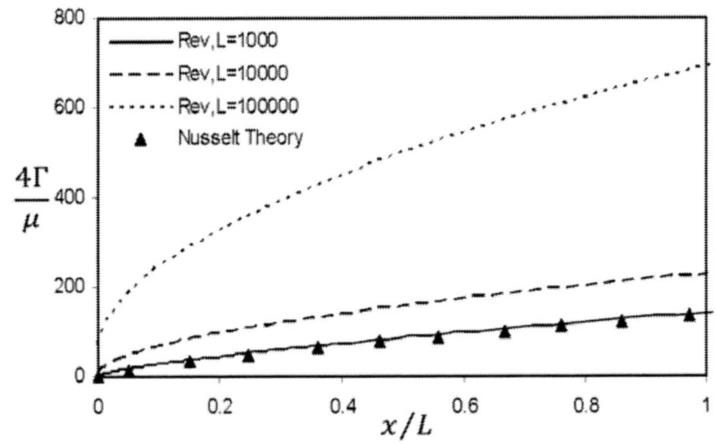

Figure 25. Variation of liquid mass flow rate per unit width (Γ) with Reynolds number.

Variation of liquid mass flow rate per unit width (Γ) for low and high vapor Reynolds numbers ($Re_{v,L}$=100000 and 1000) has been dawned in figures 26 and 27, respectively. As it has been shown in equation (89), Γ is affected by both gravitational effects of the fluid and the vapor free stream velocity. It can be seen in figure 26 that for low values of vapor Reynolds number, the mass flow rate per unit width is mostly affected by the gravitational term and the term related to vapor free stream velocity can be counted as negligible (equation (90)).

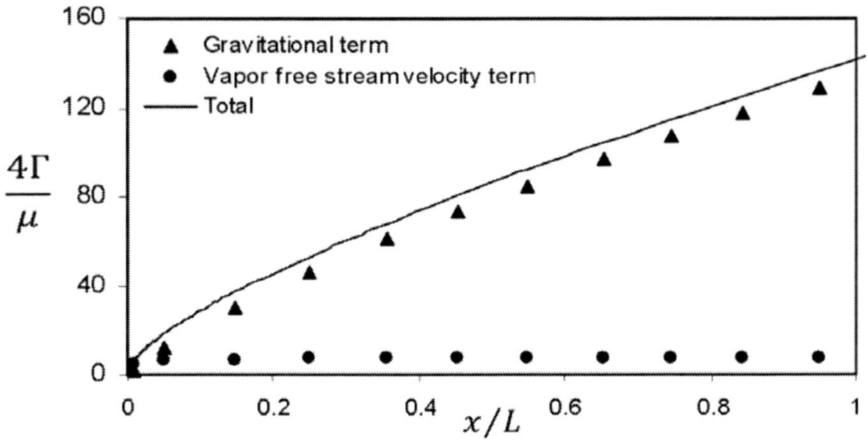

Figure 26. Variation of liquid mass flow rate per unit width (Γ) for $Re_{v,L}$=1000.

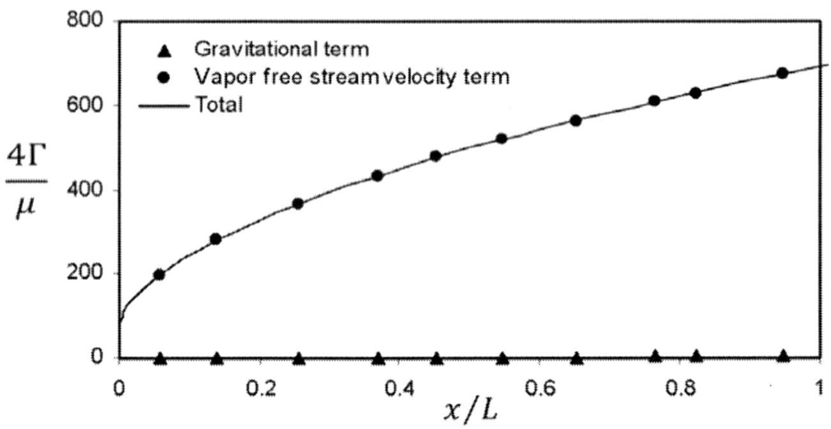

Figure 27. Variation of liquid mass flow rate per unit width (Γ) for $Re_{v,L}$=100000.

It can be seen in figure 27 that for the case of having a high vapor free stream velocity ($Re_{v,L}$=100000), the effect of the gravitational force on the value of mass flow rate is negligible compared to the effect of vapor free stream velocity. Therefore, in this case, equation (91) can be used with excellent accuracy.

So, in this section, a complete analysis on film condensation on a vertical plate with the assumption of existence a vapor boundary layer on the condensate layer was performed and the relevant results were evaluated.

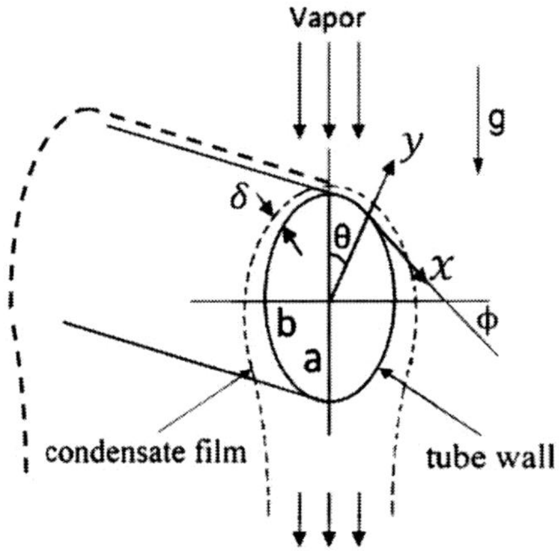

Figure 28. Physical model and coordinate system for condensate film flow on an elliptical tube.

6. Forced Convection Laminar Film Condensation on a Horizontal Elliptical Tube

In the previous sections we tried to investigate film condensation on a vertical flat plate. Now we want to consider the influences of geometry on condensation heat transfer rate. So, the problem of forced convection film condensation on an isothermal elliptical tube will be studied [68, 69]. In order to investigate the effect of geometry on the process, parameter ellipticity will be introduced and the effect of ellipticity on Nusselt number will be discussed. As stated in earlier works such as Yang and Chen [32, 79], ellipticity specifies different elliptical cylinders from $e = 0$, circular cylinder to $e = 1.0$, vertical plate.

6.1. Physical Model

Physical model and coordinate system for condensate film flow on a horizontal elliptical tube with major axis "2a" in the gravitational direction and minor axis "2b" are shown in figure 28. Outer surface of the tube is subjected to a downward flowing pure vapor that is at its saturated temperature T_{sat} and moves at uniform velocity u_{∞}. The uniform wall temperature T_w is below the saturation temperature. Thus, vapor condenses on the surface of the tube and forming a thin film of liquid that flows down the tube surface under the influence of gravity. The thickness of the film, δ, and the local mass flow rate increase with distance down the tube as condensate forms continuously along the entire liquid-vapor interface.

6.2. Mathematical Model

According to the proposed physical model, the boundary layer equations governed by the basis of conservation of mass, momentum, and energy for a laminar, steady state condensate film with constant fluid properties are respectively as:

$$\frac{\partial u}{\partial x}+\frac{\partial v}{\partial y}=0 \tag{95}$$

$$\mu\frac{\partial^2 u}{\partial y^2}=\frac{dP}{dx}-(\rho-\rho_\upsilon)g(\sin\phi) \tag{96}$$

$$k\frac{\partial^2 T}{\partial y^2}=0 \tag{97}$$

where $\phi=\phi(x)$ is the angle between the tangent to the tube wall and the horizontal direction at position $(r,\ \theta)$. The difference between equations (96) and (40) is in the term of pressure gradient which has appeared in equation (96) due to consider the effect of geometry while this term was omitted in the problem of condensation on a flat plate (equation (40)). As it seen the two other governing equations (equations (95) and (97)) are similar to those which were introduced for condensation on a vertical plate.

The related boundary conditions are:

$$y=\delta\,;\quad \frac{\partial u}{\partial y}=\frac{\tau}{\mu};\quad T=T_{Sat} \tag{98}$$

$$y=0\,;\quad u=0;\quad T=T_w \tag{99}$$

which are similar to the boundary conditions defined in sections 4 and 5, for condensation a vertical flat plate.

In writing the conservation equations, inertia effect, viscous dissipation, and axial conduction were neglected. Furthermore, it was assumed that the condensate film thickness is much smaller than the curvature diameter.

Additionally, an energy balance can be written for an element of the condensate film of height δ and width dx that was introduced before in equation (48) except that here the modified latent heat will be used as:

$$h_{fg}^{'} m^{''} = k\frac{\partial T}{\partial y} = k\frac{\Delta T}{\delta} \tag{100}$$

where $h_{fg}^{'}$ that was defined before in equation (1), is the latent heat of condensation that consider the effect of condensate sub-cooling pioneered by Rohsenow [2]. The local mass flow rate can be obtained as (which was defined before in equation (50)):

$$m^{''} = \rho\frac{d}{dx}\int_0^{\delta} u dy \tag{101}$$

By assuming that film thickness is very thin compared with the radius of surface curvature, the pressure gradient which has been appeared in momentum equation can be approximately expressed by Bernoulli‘s equation:

$$\frac{dP}{dx} = -\rho_v u_e \frac{du_e}{dx} \tag{102}$$

Next applying potential flow theory to a flow with uniform velocity u_∞, the vapor velocity at the edge of the boundary layer can be derived as [24]:

$$u_e = u_\infty\left(1 + \sqrt{(1 - e^2)}\right)\sin(\phi) \tag{103}$$

where $e = \frac{\sqrt{a^2 - b^2}}{a}$ and is the ellipticity of ellipse. It should be noted that ellipticity is different from eccentricity which is defined as b/a. The Chain derivative should be used to compute velocity gradient. By inserting u_e into pressure gradient and then substituting the pressure gradient into equation (96), the final momentum equation can be gained as follows:

$$\frac{\partial^2 u}{\partial y^2} = -(\rho - \rho_\upsilon)g(\sin\phi) + \left(-\frac{\rho_\upsilon u_\infty{}^2 \sin(2\phi)\left(1+\sqrt{(1-e^2)}\right)^2 \sqrt{(1-e^2\sin^2(\phi))^3}}{2a(1-e^2)} \right) \quad (104)$$

The vapor shear which has been used in boundary condition should be modeled. A good approximation for high condensate rates is given by the Shekriladze and Gomelauri model [16] and it will be used in the present study. This method has more usefulness and simplicity than a numerical approach, as mentioned in Jacobi [80].

$$\tau = m'' u_e = m'' u_\infty \left(1 + \sqrt{(1-e^2)}\right)\sin(\phi) \quad (105)$$

By integrating equation (97) and (104) with respect to the boundary conditions the following velocity and temperature profiles will be achieved as:

$$u = \frac{(\rho-\rho_\upsilon)g}{\mu}(\sin\phi)\,\delta^2\left(\frac{y}{\delta} - \frac{1}{2}\left(\frac{y}{\delta}\right)^2\right) + \frac{1}{2\mu}\frac{\rho_\upsilon u_\infty{}^2\left(1+\sqrt{(1-e^2)}\right)^2\sqrt{(1-e^2\sin^2(\phi))^3}}{a(1-e^2)} \times$$

$$\sin(2\phi)\,\delta^2\left(\frac{y}{\delta} - \frac{1}{2}\left(\frac{y}{\delta}\right)^2\right) + \frac{m'' u_\infty\left(1+\sqrt{(1-e^2)}\right)}{\mu}\sin(\phi) \times y \quad (106)$$

$$T(y) = (T_{sat} - T_w)\frac{y}{\delta} + T_w = \Delta T\,\frac{y}{\delta} + T_w \quad (107)$$

From equation (100), m'' can be derived as:

$$m'' = \frac{k}{h'_{fg}}\frac{\Delta T}{\delta} \quad (108)$$

Equation (108) is identical to equation (101), so m'' will be eliminated. By substituting velocity profile into equation (101) and integrating the updating equation, also introducing the dimensionless parameter, the following relation will be resulted:

$$2\delta^* \frac{d}{d\phi}\left[\frac{F}{3}\delta^{*3}(\sin\phi) + \frac{P}{6}\delta^{*3} D_e \frac{\left(1+\sqrt{(1-e^2)}\right)^2 \sin(2\phi)\sqrt{(1-e^2\sin^2(\phi))^3}}{a(1-e^2)} + \right.$$

12δ∗1+1−e2sinϕ=2Dea1−e21−e2sin2ϕ3

(109)

Due to the symmetry of the problem, the initial condition is given by:

$$\frac{d\delta^*}{d\phi} = 0 \qquad \text{at } \phi = 0 \tag{110}$$

where

$$\delta^* = \frac{\sqrt{Re}}{D_e}\delta, F = \frac{\frac{Ar_{D_e}}{Ja}}{Re^2}, P = \frac{\rho_\upsilon}{\rho}\frac{Pr}{Ja}$$
$$Pr = \frac{\mu C_p}{k}, \quad Ja = \frac{C_p \Delta T}{h'_{fg}}, \quad Re = \frac{\rho u_\infty D_e}{\mu} \tag{111}$$

and Ar_{D_e} has been introduced before in equation (47) but this time the characteristic length is the equivalent diameter of ellipsoid which is defined as:

$$D_e = 2\frac{a}{\pi}\int_0^\pi \frac{1-e^2}{\sqrt{(1-e^2\sin^2\phi)^3}}d\phi \tag{112}$$

As it has been concluded in previous works such as [31] and [81], according to F, two regions can be introduced. For high values of F or in other words when $F \to \infty$, it is said that free convection region is dominant and no effect of vapor shear is considered. Furthermore, for low F ($F \to 0$), the region is called forced convection and the effect of gravity is neglected. So the influence of free or forced convection is expressed by the value of F parameter. In this paper, the attention is on forced convection film condensation and F=0.1 has been adopted. Moreover, the pressure gradient term, P, is attributable to potential flow and practical values of this dimensionless parameter is assumed to be in the range of 0.01-1(0.01< P <1). Consequently P=1 has been chosen for this study.

6.3. Film Thickness

By applying the condition from equation (110) into equation (109), an expression for condensate film thickness at $\phi = 0$, δ^*_0, can be obtained. Thus, the film thickness can be approached by applying a numerical method of fourth-order Runge-Kutta integration.

6.4. Nusselt Number

As in Nusselt theory [1], a local heat transfer coefficient can be given as:

$$h = \frac{k}{\delta} \tag{113}$$

therefore, the dimensionless local heat transfer coefficient can be expressed as:

$$\mathrm{Nu} = \frac{hD_e}{k} = \frac{D_e}{\delta} = \frac{\sqrt{\mathrm{Re}}}{\delta^*} \tag{114}$$

in which, δ has been substituted according to δ^* from equation (111).

6.5. Results and Discussion

It has been proved that in the forced convection region, the higher the ellipticity, e, is, the larger the critical angle is [31]. Critical angle relates to the angle at which separation occurs. So by increasing the ellipticity, the location of the separation shifts more downstream for the elliptic tube. Consequently, In order to compare different parameters for various ellipticities, the range of $0 < \phi < \frac{\pi}{2}$ has been chosen.

In figure 29 variation of non-dimensional thickness of condensate layer via ϕ, for various eccentricities has been plotted. It demonstrates that for a specified eccentricity, film thickness increases as ϕ increases until it separates from the surface of the tube. Also, the figure shows that how the parameter eccentricity influences the condensate layer thickness. It can be seen that for the front portion of the tube, decreasing eccentricity will cause reduction in the value of film thickness but this trend is not true for the other section of the tube surface. Moreover, in figure 30, the curves display that how ellipticity of ellipse influences δ^* at each ϕ. It can be concluded that for low values of ϕ, increasing ellipticity decreases the film thickness while at $\phi = \pi/2$, by increasing ellipticity, δ^* has increased.

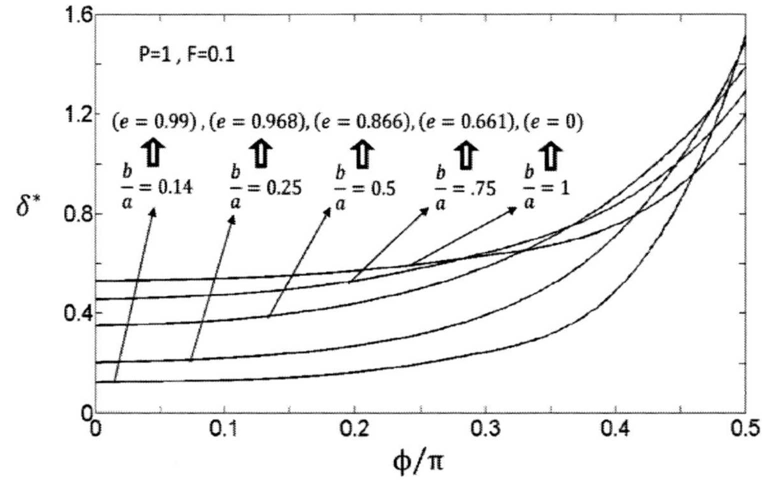

Figure 29. Variation of film thickness versus ϕ for various eccentricities.

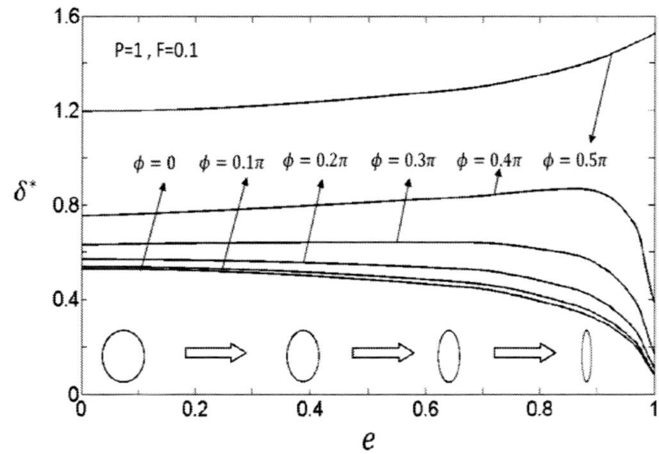

Figure 30. Variation of film thickness versus ellipticity for some selected ϕ.

Variation of local Nusselt number versus ϕ for a specified eccentricity has been plotted in figure 31. As it can be observed, Nu decreases as ϕ increases. This result is due to the increscent in film thickness as condensate layer moves downward the tube. Also, the effect of Reynolds on Nusselt number has been investigated. The figure shows that increasing Re cause increscent in the values of Nu. This result was expected since it has been proved in equation (114) that Nu is proportional to Reynolds number. So, film condensation heat transfer in forced convection can be increased by increasing the value of Reynolds number.

The effect of eccentricity on local Nusselt number has been depicted in figure 32. It was seen before in figure 29 that decreasing eccentricity affects the condensate thickness and also it was concluded in figure 31 that the amount of Nu depends on the variation of δ^*, thus as it can be seen in this figure, variation of eccentricity influences Nu number too. As it was mentioned before, for the front portion of the tube, decreasing eccentricity will cause reduction in the value of film thickness but this trend is not true for the other section of the tube surface. Figure 32 shows the variation of Nu with decreasing eccentricity is contrary to the variation of δ^*. In other word wherever condensate thickness slakes down by decreasing eccentricity, Nu number increases. This trend is due to the reverse proportion of Nu to δ^* which was shown before in equation (114).

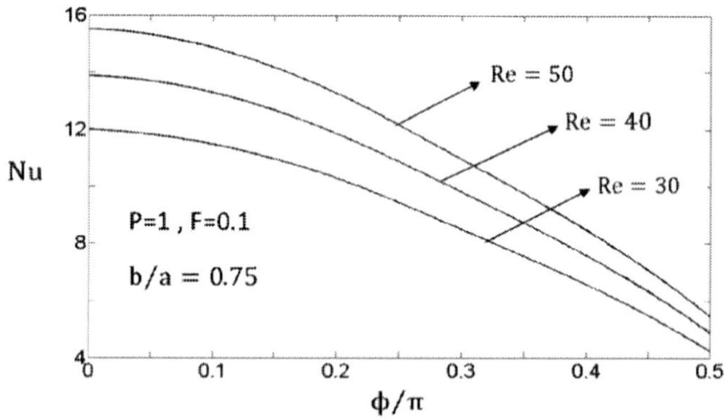

Figure 31. Variation of local Nusselt number versus ϕ for different Reynolds number.

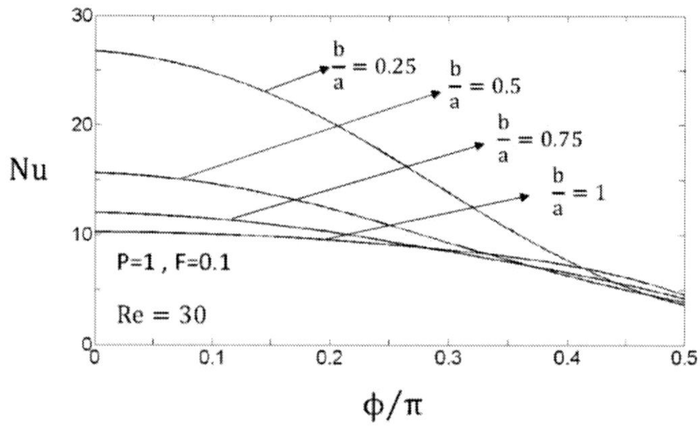

Figure 32. Variation of Nusselt number versus ϕ for different eccentricities.

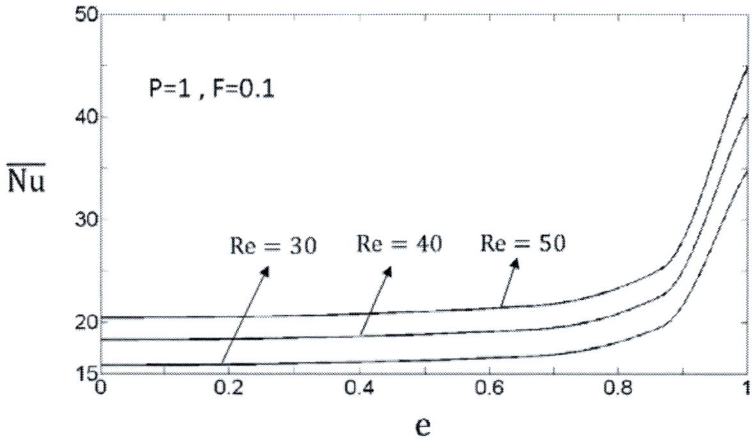

Figure 33. Variation of average Nusselt number versus ellipticity of ellipse for different Re number.

Variation of average Nusselt number, $\overline{Nu}$, which can be obtained by integrating Nu with respect to ϕ from zero to $\pi/2$, versus different ellipticity of ellipse has been plotted in figure 33. It can be concluded from this figure that the effect of ellipticity on average Nusselt number is significant for $e > 0.7$ and for this range of ellipticity, increasing ellipticity, increases the amount of average Nusselt number. Also, the figure confirmed that Reynolds number influences $\overline{Nu}$ and as Re increases, $\overline{Nu}$ increases.

So, through this analysis, we tried to study the problem of film condensation on a surface different from the flat plate.

CONCLUSION

In this chapter, we tried first to introduce more details of condensation heat transfer on a vertical plate without interfacial shear stress. A computer program has been developed to solve the problem. A heat transfer coefficient of a condensate film for a wide range of physical properties has been obtained and presented graphically and also has been discussed. In the next step we have considered shear stress at the liquid-vapor interface and analytically solved the problem of condensation on an isothermal plate. The solution was based on categorizing two limited cases as condensation with high interfacial shear force

($Ri_x \rightarrow \infty$) and condensation with no interfacial shear force ($Ri_x \rightarrow 0$, Nusselt theory).

Also, we have studied analytically the phenomenon of condensation with assisted flowing vapor on isothermal flat plate. Eventually, in order to consider the influences of geometry on laminar condensation heat transfer rate, an analysis on forced convection film condensation on an elliptical tube has been done and the effect of Reynolds and ellipticity of ellipse on Nusselt number have been discussed.

The conclusions obtained through the chapter can be classified as:

- Laminar film condensation on a vertical plate without interfacial shear stress
  - The Nusselt theory is acceptable for low Jacob numbers at all Prandtl numbers.
  - As Jacob number increases, the results diverge from the Nusselt theory.
  - Divergences from Nusselt theory are from level one downward at low *Pr* and upward for high *Pr*.
  - Results for high *Pr* ($Pr=100$) are usable for $Pr/Ja<0.1$.
  - Results for low *Pr* ($Pr=0.01$) are usable for low *Ja* ($Ja<0.1$).
- Laminar film condensation on a vertical plate with assisted interfacial shear stress
  - By considering shear stress, the velocity profile has greater values in the cases with higher interfacial forces compared to the Nusselt case were no shear stress is assumed to exert upon the condensate layer.
  - For the presence of shear stress at the interface of liquid-vapor, in both categories which were defined as $Ri_x \rightarrow \infty$ and $Ri_x \rightarrow 0$, an analytical solution can be achieved for the thickness of the condensate layer as well as the Nusselt number which are proved to be functions of two dimensionless groups: $4Ja/\Pr Ar_x^2$ and $1.5Ri_x$.
  - In condensation assisted with shear stress, for very low values of Richardson number ($Ri_x< 0.1$), the analytical limiting solution for the case of $Ri_x<<1$, and for very high values of Richardson number ($Ri_x>10$), the analytical limiting solution for the first case ($Ri_x>>1$) can be taken with excellent degree of approximation (error< 5%). These Ranges can be extended to $Ri_x<<0.2$ and $Ri_x>>1.6$ with the error of 15% near $Ri_x=1.6$.

- In the calculations of liquid film thickness and the Nusselt number for the case of considering the effect of shear stress in condensation analysis, an approximate solution with an error of less than 3% is carried out for $0.2 \leq Ri_x \leq 2.5$ where the gravitational and interfacial shear forces have comparable values.
- The interfacial shear forces will result in an increase in the liquid mass flow rate.

- Laminar film condensation on a vertical plate with assisted vapor flow
  - The velocity profile has greater values in the cases with higher values of vapor free stream velocity compared to the Nusselt case where the vapor free stream velocity is assumed to be zero and no shear stress is assumed to exert upon the condensate layer.
  - By assuming constant properties for the liquid (constant ν), it has been resulted that the velocity profile has greater values in the cases with higher vapor velocities compared to the Nusselt case.
  - For condensation with assisted vapor flow, the values of the interfacial velocity are very low compared to the vapor free stream velocity (about 1%).
  - When an associated vapor flow is forced past a cool wall, it will drag the liquid film along and thinning it out.
  - Thinner liquid film thickness enhances the heat transfer coefficient.
  - In condensation with assisted vapor flow, higher values of *Ja* will lead to higher values of the dimensionless liquid film thickness and consequently a lower heat transfer coefficient.
  - The vapor free stream velocity will result in an increase in the liquid mass flow rate.
  - For constant values of $Re_{v,L}$ number in condensation assisted with vapor flow, increasing the Ar_L number will result in an increase in the *Nu* number.
- Forced convection laminar film condensation on a horizontal elliptical tube
  - In condensation on the outer surface of elliptical tube, increasing ellipticity or decreasing eccentricity, influences the condensate film thickness which was explained in figure 29.
  - By increasing ϕ or as condensate layer moves downward the tube, Nusselt number decreases.

- Variation of Nusselt number by varying eccentricity in condensation on the elliptical tube is contrary to the variation of condensate thickness by changing ellipticity.
- Increasing Re causes increscent in the values of Nusselt number. So, film condensation heat transfer in forced convection film condensation on the elliptical tube can be increased by increasing the value of Reynolds number.
- For an elliptical tube encountered in condensation process, effect of ellipticity on $\overline{Nu}$ is significant for e >0.7 and for this range of ellipticity, increasing e, increases the amount of $\overline{Nu}$.

The results of the studies which were reviewed in this section can be applied in any work of external laminar film-wise condensation with the mentioned conditions.

REFERENCES

[1] Nusselt, W. 1916. Dia Oberflachenkondensation des wasserdampfes. *Zeitschrift des Vereines deutscher Ingenieure*. Vol.60, 1645-1648.

[2] Rohsenow, W.M. 1916. Heat transfer and temperature distribution in laminar film condensation. *TRANS. ASME*. Vol.78, 541-569.

[3] Sparrow, E.M; Gregg, J.L. 1959. A boundary-layer treatment of laminar film condensation. International *Journal of Heat Transfer*. Vol.81, 13-18.

[4] Merte, Jr.H. 1973. Condensation heat transfer. *J. of Adv. Heat Transfer*. Vol.15, 181-272.

[5] Koh, J.C.Y. 1961. On integral treatment of two-phase boundary layer in film condensation. *Journal of Heat Transfer*. Vol.83, 48-54.

[6] Chen, M.M. 1961. An analytical study of laminar film condensation. *Journal of Heat Transfer*. Vol.83, 48-54.

[7] Tanasawa, I. 1991. Advances in condensation heat transfer. J. *of Adv. Heat transfer*. Vol.21, 55-139.

[8] Koh, J.C.Y. 1962. Film condensation in a forced convection boundary layer flow. *International Journal of Heat and Mass Transfer*. Vol.5, 941-954.

[9] Chen, S.L; Gerner, M.F; Tien, C.L. 1987. General film condensation correlations. *Journal of Experimental Heat Transfer*. Vol.1, 93-107.

[10] Mendez, F; Lizard, J.J; Trevino, C. 2000. Laminar film condensation along a vertical fin. *International Journal of Heat and Mass Transfer*. Vol.43, 2859-2868.

[11] Xu, H; You, X.C; Pop, I. 2008. Analytical approximation for laminar film condensation of saturated stream on an isothermal vertical plate. *Applied Mathematical Modeling*. Vol.32, 738-748.

[12] Fujii, T; Uehara, H. 1972. Laminar film-wise condensation on a vertical surface. *International of Heat and Mass Transfer*. Vol.15, 217-233.

[13] Poots, G; Miles, R.G. 1967. Effects of variable physical properties on laminar film condensation of saturated steam on a vertical flat plate. *International Journal of Heat and Mass Transfer*. Vol.10, 1677-1692.

[14] Koh, J.C.Y; Sparrow, E.M; Harnett, J.P. 1961. The two phase boundary layer in laminar film condensation. *International Journal of Heat and Mass Transfer*. Vol.2, 69-82.

[15] Sugawara, S; Michiyoshi, I; Minamiyama, T. 1956. The.condensation of vapor flowing normal to a horizontal pipe. *Proc Sixth Jap Nat Cong App Mech paper III*. Vol.4, 385-388.

[16] Shekriladze, L.G; Gomelauri, V.I. 1966. Theoretical study of laminar film condensation of a flowing vapor. *International Journal of Heat and Mass Transfer*. Vol.9, 581-591.

[17] Hsu, C.H; Yang, S.A. 1999. Pressure gradient and variable wall temperature effects during film-wise condensation from downward flowing vapors onto a horizontal tube. *International Journal of Heat and Mass Transfe*r. Vol.42, 2419-2426.

[18] Rose. J.W. 1984. Effect of pressure gradient in forced convection film condensation o a horizontal tube-effect of surface temperature variation. *International Journal of Heat and Mass Transfer*. Vol.27, 39-47.

[19] Yang, S.A; Chen, C.K. 1992. Role of surface tension and ellipticity in laminar film condensation on a horizontal elliptical tube. *International Journal of Heat and Mass Transfer*. Vol.36, 3135-3141.

[20] Rose, J.W. 1998. Condensation heat transfer fundamentals. *Trans IChemE: Part* A. Vol.76, 143-152.

[21] Rose, J.W. 2004. Surface tension effects and enhancement of condensation heat transfer. *Trans. IChemE: Part A*. Vol.82, 419-429.

[22] Kibboua, R; Azzi, A. 2005. Laminar film condensation on an elliptical tube embedded in porous media. *Energy Conversion and Management*. Vol.46, 2359-2372.

[23] Chang, T.B; Yeh, W.Y. 2009. Effects of uniform suction and surface tension on laminar film wise condensation on a horizontal elliptical tube in a porous medium. *International Journal of Thermal Sciences*. Vol.48, 2323-2330.

[24] Winkler, C.M; Chen, T.S; Minkoweycz, W.J. 2001. Film condensation of saturated and superheated vapors along isothermal vertical surfaces in mixed convection. *Numerical Heat Transfer Part A: Applications*. Vol.36, 375-393.

[25] Yang, S.A; Hsu, C.H. 1997. Mixed convection film condensation on a horizontal elliptical tube with uniform surface heat flux. *Numerical Heat Transfer Part A: Applications*. Vol.32, 85-95.

[26] Al-Jarrah, J.A; Khadwari, A.F; Nimr, M.A. 2008. Film condensation on a vertical micro-channel. *International Communications in Heat and Mass Transfer*. Vol.35, 1172-1176.

[27] Honda, H; Fujii, T. 1984. Condensation of a flowing vapour on a horizontal tube-numerical analysis as a conjugate heat transfer problem. *Journal of Heat Transfer*. Vol.106, 841-848.

[28] Memory, S.B; Lee, W.C; Rose, J.W. 1993. Forced convection film condensation on a horizontal tube-effect of surface temperature variation. *International Journal of Heat and Mass Transfer*. Vol.38, 1671-1676.

[29] Memory. S.B; Lee, W.C; Rose, J.W. 1995. Forced convection film condensation on a horizontal tube-influence of vapor boundary layer separation. *International Journal of Heat and Mass Transfer*.Vol.117, 529-533.

[30] Michael, A.G; Rose, J.W; Daniels, L.C. 1989. Forced convection condensation on a horizontal tube-effect of experiments with vertical down-flow of steam. *ASME*. Vol.111, 792-797.

[31] Yang, S.A; Hsu, A.H. 1997. Free and forced convection film condensation from a horizontal elliptical tube with a vertical plate and horizontal tube as special cases. *International Journal of Heat and Fluid Flow*. Vol.18, 567-574.

[32] Yang, S.A; Chen, C.K. 1993. Role of surface tension and ellipticity in laminar film condensation on horizontal elliptical tube. *International Journal of Heat and Mass Transfer*. Vol.36, 3135-3141.

[33] Ali, A.F.M; McDonal, T.W. 1977. Laminar film condensation on horizontal elliptical cylinders: A first approximation for condensation on inclined tubes. *ASHRAE Trans*. Vol.83, 242-249.

[34] Karimi, A. 1977. Laminar film condensation on helical reflux condensers and related configurations. *International Journal of Heat and Mass Transfer*. Vol.20, 1137-1144.

[35] Memory, S.B; Rose, J.W. 1991. Free convection laminar film condensation on a horizontal tube with variable wall temperature. *International Journal of Heat and Mass Transfer*. Vol.34, 2775-2778.

[36] Chang, T.B; Yeh, W.Y. 2010. Theoretical investigation into condensation heat transfer on horizontal elliptical tube in stationary saturated vapor with wall suction. *Applied Thermal Engineering*. 1-8.

[37] Dhir, V.K; Linhard, J.H. 1973. Similar solutions for laminar film condensation with variable gravity or body shape. *ASME Journal of Heat Transfer*. Vol.95, 97-100.

[38] Cess, R.D. 1960. Laminar film condensation on a flat plate in the absence of a body force. *Z Angew Math Phys*. Vol.11, 426-433.

[39] Jacobs, H.R. 1966. An integral treatment of combined body force and forced convection in a laminar film condensation. *International Journal of Heat and Mass Transfer*. Vol.9, 637-648.

[40] Churchill, S.W. 1986. Laminar film condensation. *International Journal of Heat and Mass Transfer*. Vol.29, 1219-1226.

[41] Mayhew, Y.R; Griffiths, D.J; Philips, J.W. 1965. Effect of vapor drag on laminar film condensation on a vertical surface. *Proc I Mech E*. Vol.180, 280-287.

[42] Vorskresenskiy. 1948. Calculation of heat transfer in film condensation allowing for the temperature dependence of the physical properties of the condensate. *U.S.S.R. Acad. Sci*.

[43] Labuntsov, D.A. 1957. Effect of temperature dependence of physical parameters of condensate on heat transfer in film condensation of steam. *Teploenergetika*. Vol.4, 49-51.

[44] Brouwers, H.J.H. 1988. Film condensation on non-isothermal vertical plates. *International Journal of Heat and Mass Transfer*. Vol.32, 655-663.

[45] Shang, D.Y; Adamek, T. 1994. Study on laminar film condensation of saturated steam on a vertical flat plate for condensation of various physical factors including variable thermo-physical properties. *Warme. Und Stoffubertragung*. Vol.30, 89-100.

[46] Karabelas, S.J; Markatos, N.C. 2008. Water vapor condensation in forced convection flow over an airfoil. *Aerospace Science and Technology*. Vol.12, 150-158.

[47] Dengliang, Z; Hong, X; Baojin, Q. 2010. Numerical heat transfer analysis of laminar film condensation on a vertical fluted tube. *Applied Thermal Engineering*. Vol.30, 1159-1163.

[48] Baojin, Q; Li, Z; Hong, X; Yan, S. 2011. Experimental study on condensation heat transfer of steam on vertical titanium plates with different surface energies. *Experimental Thermal and Fluid Science*. Vol.35, 211-218.

[49] Monning, C; Numrich, R. 1999. Condensation of vapours of immiscible liquids in the presence of a non-condensable gas. *International Journal of Thermal Science*. Vol.38, 684-694.

[50] Al-Diwany, H.K; Rose, J.W. 1973. Free convection film condensation of steam in the presence of non-condensable gases. *International Journal of Heat and Mass Transfer*. Vol.16, 1359-1369.

[51] Minkowycz, W.J; Sparrow, E.M. 1966. Condensation heat transfer in the presence of non-condensables, interfacial resistance, super heating, variable properties, and diffusion. *International Journal of Heat and Mass Transfer*. Vol.9, 1125-1144.

[52] Denny, V.E; Jusionis, V.J. 1972. Effects of noncondensable gas and forced flow on laminar film condensation. *International Journal of Heat and Mass Transfer*. Vol.15, 315-326.

[53] Denny, V.E; Mills, A.F, Jusionis, V.J. 1971. Laminar film condensation from a steam-air mixture undergoing forced flow down a vertical surface. *Journal of Heat Transfer*. Vol.93, 297-304.

[54] Denny, V.E; South III, V. 1972. Effects of forced flow, noncondensables, and variable properties on film condensation of pure and binary vapors at the forward stagnation point of a horizontal cylinder. *International Journal of Heat and Mass Transfer*. Vol.15, 2133-2142.

[55] Ganguli, A; Patel, A.G; Maheshwari, N.K; Pandit, A.B. 2008. Theoretical modeling of condensation of steam outside different vertical geometries (tube, flat plates) in the presence of non-condensable gases like air and helium. *Nuclear Engineering and Design*. Vol.238, 2328-2340.

[56] Chen, C.K; Lin, Y.T. 2009. Laminar film condensation from a downward flowing steam-air mixture onto a horizontal circular tube. *Applied Mathematical Modeling*. Vol.33, 1944-1956.

[57] Chen, C.K; Lin, Y.T. 2009. Turbulent film condensation in the presence of non-condensable gases over a horizontal tube. *International Journal of Thermal Sciences*. Vol.48, 1777-1785.

[58] De La Rosa, J.C; Escriva, A; Herranz, L.E; Cicero, T; Munoz-Cobo, J.L. 2009. Review on condensation on the containment structures. *Progress in Nuclear Energy*. Vol.51, 32-66.

[59] Adeyinka, O.B; Naterer, G.F. 2004. Optimization for entropy production and energy availability in film condensation. *Int. Commun. Heat Mass Transfer 31*. Vol.4, 513-524.

[60] Yang, S.A; Guan, C.L; Wen, J.Y. 2007. Thermodynamic optimization of free convection film condensation on a horizontal elliptical tube with

variable wall temperature. *International Journal of Heat and Mass Transfer*. Vol.50, 4607-4613.

[61] Dung, S.C; Yang, S.A. 2006. Second law based optimization of free condensation film wise condensation on horizontal tube. International *Communications in Heat and Mass Transfer*. Vol.33, 636-644.

[62] Li, G.C; Yang, S.A. 2007. Entropy generation minimization of free convection film condensation on an elliptical cylinder. *International Journal of Thermal Sciences*. Vol.46, 407-412.

[63] Yang, S.A; Li, G.C; Yang, W.J. 2007. Thermodynamic optimization of free convection film condensation on a horizontal elliptical tube with variable wall temperature. *International Journal of Heat and Mass Transfer*. Vol.50, 4607-4613.

[64] Li, G.C; Yang, S.A. 2006. Thermodynamic analysis of free convection film condensation on an elliptical cylinder. *Journal of the Chinese Institute of Engineers*. Vol.29, 903-908.

[65] Chen, B.C; Wu, J.R; Yang, S.A. 2008. Entropy generation of forced convection film condensation from downward flowing vapors onto a horizontal tube. *Journal of the Chinese Institute of Engineers*. Vol.31, 349-354.

[66] Esfahani, J.A; Koohi, S. 2008. Entropy generation minimization in film condensation on a flat plate with interfacial shear stress. *The 16th Annual (International) Conference on Mechanical Engineering-ISME.*

[67] Esfahani, J.A; Koohi-Fayegh, S. 2009. Entropy generation analysis in film condensation on an elliptical tube with interfacial shear stress at high Bond numbers. *5th Dubrovnik Conference on Sustainable Development of Energy Water and Environmental Systems.*

[68] Esfahani, J.A; Modirkhazeni, S.M. 2011. Effect of Brinkman number on entropy generation of forced convection film condensation on the outer surface of elliptical tube. *1st Iranian Thermal Science Conference-ITSC.*

[69] Esfahani, J.A; Modirkhazeni, S.M. 2011. Effect of ellipticity on entropy generation of forced convection film condensation on the outer surface of tube. *The 19th Annual (International) Conference on Mechanical Engineering-ISME.*

[70] Esfahani, J.A; Ziaei-Rad, M. 2002. Analysis of laminar film condensation on a vertical plate. *The Ninth Asian Congress of Fluid Mechanics.*

[71] Esfahani, J.A; Ziaei-Rad, M. 2005. Study of laminar film condensation on a vertical plate. *J. of Aerospace Science and Technology 2.*

[72] Esfahani, J.A; Koohi-Fayegh, S. 2009. Film condensation on a flat plate by interfacial shear stress. *I.R.E.CH.*E. Vol.1, 27-31.

[73] Esfahani, J.A; Koohi-Fayegh, S. 2009. Film condensation on a flat plate with assisted vapor flow. *Proceedings of 12th Workshop of Transport Phenomena in Two-Phase Flow.*

[74] Incropera, F.P; DeWitt, A.P. Introduction to Heat Transfer; *Sixth Edition*; Vol.10, pp 643-646.

[75] Sadasivan, P; Lienhard, J.H. 1987. Sensible heat correction in laminar film boiling and condensation. *Journal of Heat Transfer*. Vol.109, 45.

[76] William, H.P; Brain, P. F; Saul, A.T. Numerical Recipes. *Cambridge University Press.* 1986.

[77] Bejan, A. Convection heat transfer. 2nd Edition. *John Wiley and Sons*. 1995.

[78] Lock, G.S.H. 1992. The tubular thermo-siphon. *Oxford University Press*.

[79] Yang, S.A; Chen, C.K. 1993. Effects of surface tension and non-isothermal wall temperature variation upon film wise condensation on vertical ellipsoids/sphere. *Proceedings of the Royal Society of London*. Vol.442, 301-312.

[80] Jacobi, A.M. 1992. Film wise condensation from a flowing vapor onto isothermal axis symmetric bodies. *Journal of Thermo Physics and Heat Transfer*. Vol.6, 321-325.

[81] Memory, S.B; Adams, V.H; Marto, P.J. 1997. Free and forced convection laminar film condensation on horizontal elliptical tubes. *International Journal of Heat and Mass Transfer*. Vol.40, 3395-3406.

In: Water Condensation
Editor: Amimul Ahsan

ISBN: 978-1-61324-874-4

Chapter 2

WATER CONDENSATION AT NANOSCALE: AN ATOMIC FORCE MICROSCOPY STUDY

Lucel Sirghi[1]

Faculty of Physics,"Al. I. Cuza" University, bd. Carol I, Iasi, Romania

ABSTRACT

Undersaturated water vapor in air condenses in nanoscopic gaps formed between solid surfaces (i. e. microscopic contacts between macroscopic or microscopic bodies, microscopic pores on a solid surface, nanotubes, etc) and forms small water menisci. The phenomenon is known as capillary water condensation and the curvature radii of such nanoscopic water menisci at thermodynamic equilibrium with the water vapor are determined by Kelvin equation. The atomic force microscope (AFM) is the perfect tool for investigation of the phenomenon because it senses directly the small capillary force generated by a microscopic water meniscus. Usually, in such investigations the very sharp tip of the AFM probe approaches a sample surface, get in nanoscopic contact with it, and then withdraws until the contact is broken up. During the contact, the water vapor condenses forming a nanoscopic water bridge that is broke up during the tip withdrawal. The capillary force sensed by the AFM probe during the water bridge elongation and break up is a measure of the amount of water condensed at the contact.

This chapter presents the principles of atomic force microscopy investigation of capillary water condensation at nanoscopic contacts and reports the main results obtained by the author in such investigations. It is

[1]E-mail: lsirghi@uaic.ro.

shown that the amount of water condensed at a nanoscopic contact depends on the contact geometry, hydrophilic property of surfaces, surface contaminants, air humidity, and, in some cases, the contact time.

1. INTRODUCTION

For macroscopic and flat liquid-vapor interfaces the liquid phase is at thermodynamic equilibrium with the vapor one at a certain value of vapor pressure called the equilibrium vapor pressure or the saturation vapor pressure (p_0). If the vapor pressure is lower than p_0 (undersaturated vapor) the liquid evaporates while if the vapor pressure is larger than p_0 (oversaturated vapor) the vapor condenses. We may say that at p_0 the evaporation rate is equal to the condensation rate and the two phases are in thermodynamic equilibrium. The value of p_0 depends on material and temperature (p_0 increases with the increase of temperature and is larger for more volatile liquids than for less volatile liquids).

Different from macroscopic liquid-vapor interfaces, which are usually flat, the microscopic liquid-vapor interfaces (menisci) are curved and their curvatures affect the values of equilibrium vapor pressure. This is happening because molecules escape easer from a convex liquid meniscus (due to less interaction with neighbor molecules) than from a flat one while they escape harder from a concave liquid meniscus (due to stronger interaction with neighbor molecules) than from a flat one. Therefore for convex liquid menisci the vapor equilibrium pressure,p_0', is larger than p_0, while for concave menisci it is smaller than p_0. This dependence of equilibrium vapor pressure on liquid meniscus curvature is illustrated in Figure 1.

Lord Kelvin [1] was the first to notice this phenomenon and to explain its consequences for liquid condensation in small gaps, i. e. the capillary condensation. The capillary condensation is the phenomenon of condensation of undersaturated vapor ($p < p_0$) in small gaps (pores, capillary tubes and microscopic contacts between solid bodies). At certain vapor pressure, $p < p_0$, the vapor condenses in a small gap until a liquid meniscus with a certain value of the curvature radius, R_K, is build up. In the case of capillary condensation, the liquid menisci are concave and have negative curvatures, $C_K = 1/R_K < 0$. The value R_K is established by the Kelvin equation and increases in absolute value as p approaches p_0. Figure 2 shows a water liquid bridge formed by capillary condensation in the microscopic gap between two hydrophilic bodies. Such water liquid bridges form

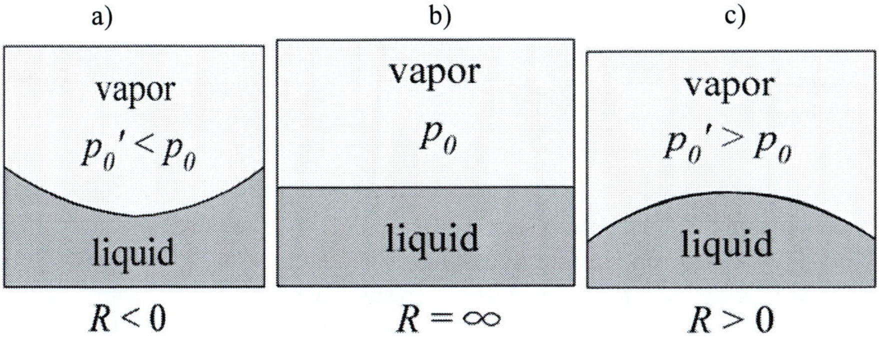

Figure 1. Dependence of vapor equilibrium pressure on curvature of liquid-vapor interface: a) for a concave meniscus, $R < 0$ and $p_0' < p_0$; b) for a flat meniscus $R = \infty$ and $p_0' = p_0$; c) for a convex meniscus, $R > 0$ and $p_0' > p_0$.

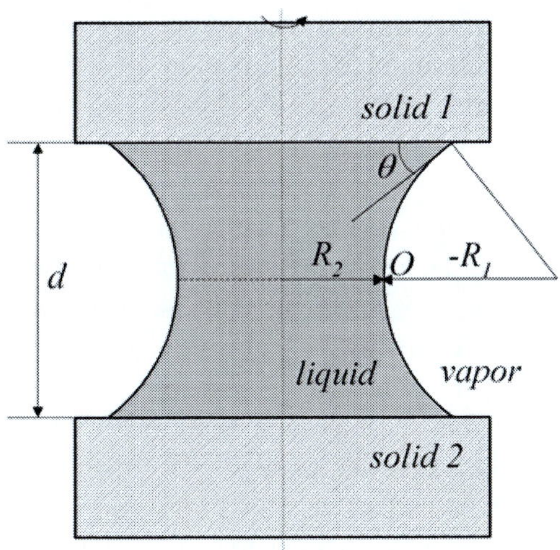

Figure 2. Formation of water bridges through capillary condensation of water vapor in a small gap (d in the range of few nanometers) between two hydrophilic solid bodies ($\theta <$ 90°) bodies. The water meniscus is concave in vertical plane and convex in horizontal plane.

in very short time after the bodies are brought at a separation distance comparable with R_K (few nanometers). For example, these microscopic water bridges are responsible for the adhesion forces between very small particles of the sand castles or in sintering process of powders.

Recently, the capillary condensation has received a great deal of attention because of its relevance to the fields of nanotribology [2], micro electro mechanic systems (MEMS) [3], colloidal physics [4], nanolithography [5] and biochemistry [6]. The capillary condensation of water affects greatly the atomic force microscopy (AFM) measurements in air [7,8]. The atomic force microscope was invented by Binnig et al. [9] in 1986 and it allows imaging of surface topography of conducting or insulating samples, in some cases with atomic resolution. The instrument is based on the detection of the interaction force between a microscopic sharpened tip and a sample surface. The attribute "atomic" comes from the fact that the interaction region between the tiny tip (radius of 10 nm) and sample includes a limited number of atoms. In the case of AFM measurements in air, capillary condensation of water vapor leads to formation of a water bridge that surrounds the nanoscopic contacts between the very sharp AFM tips and sample surfaces. Formation of water bridges by capillary condensation of water vapor in small gaps between AFM tip and sample surfaces was directly visualized by scanning electron microscopy[10]. Formation of water bridges at tip-sample contact increases greatly the tip-sample adhesion and friction forces with bad consequences for the quality of the topography images acquired during scanning of the probe surface. Capillary condensation in contact AFM can be avoided by use of vacuum or a liquid as measurement environment. However, the AFM became one of the main techniques used for studying the capillary condensation. Thus AFM techniques were used to visualize the water droplets formed by capillary condensation at tip sample contacts and their evaporation [11]. Many AFM studies have been dedicated to investigation of the effect of capillary condensation of water on adhesion and friction forces at nanocontacts of solid surfaces [12, 13]. This is also the main subject of the present chapter, which is intended to present a comprehensive view of the AFM research of water capillary condensation at nanoscopic contacts carried out by the author across the years.

2. The Atomic Force Microscope

A sketch of an atomic force microscopy (AFM) system in its most common configuration is given in Figure 3. The AFM probe, which consists of a flexible silicon or silicon nitride cantilever with a sharp tip at its end, plays the most important role in the AFM system. The force is sensed by a laser beam reflected on the top of the cantilever. Any small force acting on the AFM tip deflects the cantilever and the laser beam and the deviation (force) is measured by a photodiode. The AFM sample (in some cases, the probe) is manipulated with nanometer precision by a piezoelectric scanner. The topography image of the sample surface is acquired by raster scanning the sample surface while the tip-sample interaction force is kept constant at a set-point value by the feedback electronics of the instrument. Thus, the AFM is working as a stylus profilometer at nanoscale. The tip-sample interaction force, which is kept constant by the AFM feedback during imaging a sample surface, can be as small as 10 pN. The lateral resolution depends on the AFM tip radius and can reach few nanometers, while the vertical resolution is about 0.1 nm.

Basically, the interaction between the AFM tip and sample surfaces is the result of the superposition of van der Waals, electrostatic, capillary and interfacial tension forces [14]. The latter two forces arise due to the capillary water bridge, which usually occurs at the tip-sample contact region in the ambient air, and their contribution to the tip-sample adhesive force is dominant [15]. Therefore, in the followings the adhesion force measured by AFM pull-off experiments in air is interpreted as the capillary force, i. e. the force generated by the capillary water bridge formed at the AFM tip-sample surface.

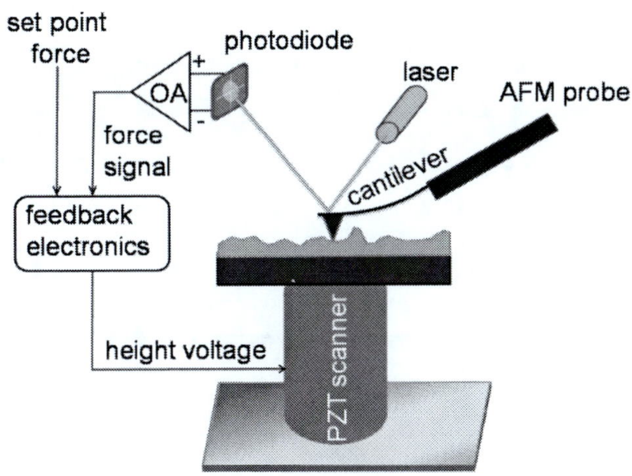

Figure 3. Principle of the atomic force microscope. The main component of the microscope is the force detector consisting in a very flexible microcantilever with its very sharp tip interacting with the sample surface and a laser beam deflected by the cantilever and recorded by a four-quadrant photodiode. The detected vertical force is used by a feedback system to control the height of the sample surface by applying the height voltage to the PZT scanner. A set-point value of the tip-sample interaction force is thus maintained constant during scanning of the probe surface.

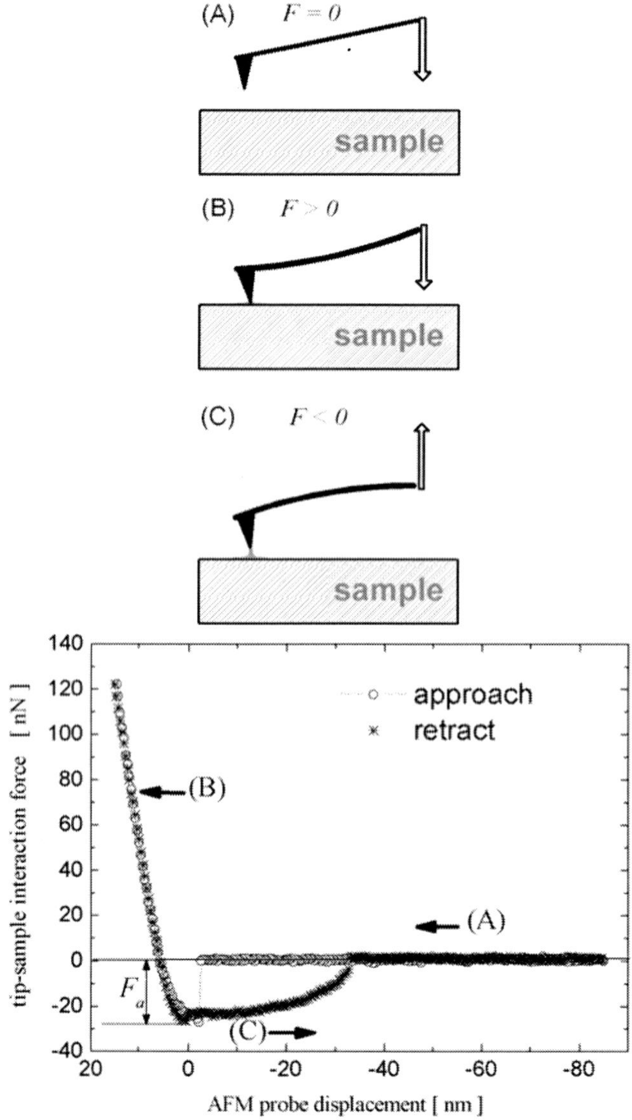

Figure 4. Typical force displacement curve recorded in an AFM pull-off experiment. *(A)* The AFM tip is approaching the sample surface while the interaction force is negligible (F=0). *(B)* The AFM tip is pushing on the sample surface. During the contact, the water vapor from the ambient air forms through capillary condensation a water neck that surrounds the tip-sample contact. *(C)* The AFM tip is retracting while the water bridge is elongated. The attraction force caused by the water bridge is the capillary force measured as the tip-sample adhesion force.

Figure 4 shows the typical variation of tip-sample interaction force in a pull-off experiment along with a sketch of the AFM probe position at three moments of time during the experiment. In such experiments the AFM tip is approached to the sample surface until contact. While the distance between the tip and sample surfaces is larger than 10 nm the tip-sample interaction force is zero (in absence of long range electric and magnetic forces). After contact, the movement of the probe continues until certain value of the repulsive interaction force (positive) is reached. Then, the movement of the AFM probe is reversed and it is retracted until the initial position. The maximum attractive force (negative) measured during tip retraction is considered as the tip-sample adhesive force.

3. Theory of Capillary Condensation

Understanding of capillary condensation phenomenon is based on two important thermodynamic equations of liquid-vapor interfaces: the Young-Laplace equation and the Kelvin equation [16]. In the followings, these two equations are derived and their consequences are discussed.

3.1. Young-Laplace Equation

The Young-Laplace equation concerns the relationship between pressure values on the two sides of a curved liquid-vapor interface. It states that the pressure of liquid, p_l, differs the pressure of vapor, p_v, with the quantity γ/R, where γ is the superficial tension of the liquid and R, the curvature radius of the meniscus.

$$p_l - p_V = \frac{\gamma}{R} \tag{1}$$

Here, the curvature radius is positive for convex menisci and negative for concave menisci. Therefore, for a convex meniscus $p_l > p_V$, while for a concave meniscus $p_l < p_V$. The superficial tension of liquid, γ, is the energy required to increase its interface with vapor by an area unit:

$$\gamma = \frac{dW}{dA} \tag{2}$$

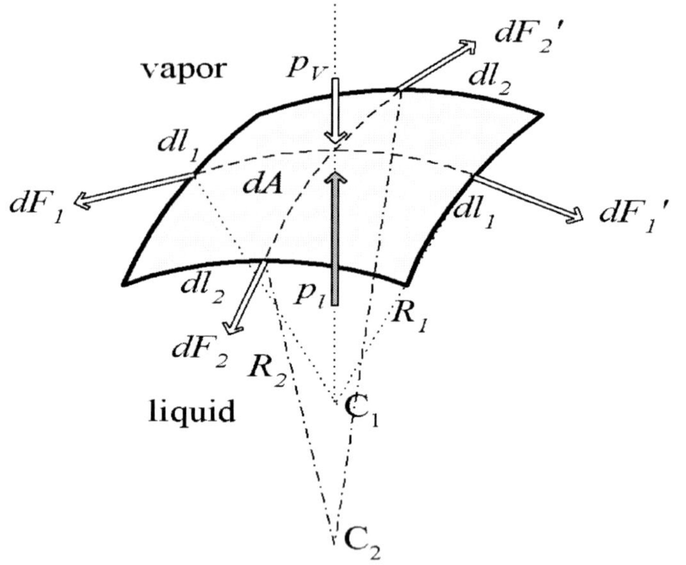

Figure 5. Forces that act on an elementary surface *dA* as result of superficial tension and liquid and vapor pressures.

Since the process of surface variation takes place usually at constant pressure, temperature, and number of molecules, the energy *dW* is equal to Gibs free energy, *dG*. Thus, γ is defined as:

$$\gamma = \left.\frac{dG}{dA}\right|_{p,T,N} \tag{3}$$

Derivation of Young-Laplace equation considers the mechanical equilibrium of forces acting on the liquid-vapor interface. On a small square element of a meniscus surface of area, $dA = dl^2$, acts the vapor force $p_V dA$ and the liquid force $p_l dA$. On the four contour lines of *dS* (Figure 2) act four forces of intensity $\gamma \cdot dl$. For a planar interface ($R = \infty$) these four superficial tension forces are in the plane of interface and therefore they cancel each other. In this case, the balance of forces acting on the two sides of the meniscus requires that $p_l = p_V$. For the case of convex meniscus the superficial tension forces are not acting in plane and, therefore, p_l should be larger than p_V in order to maintain the mechanical equilibrium of interface. In this case forces γdl have the vertical components $\gamma dl \cdot \sin(d\alpha/2)$ and $\gamma dl \cdot \sin(d\beta/2)$, respectively, acting towards liquid. Here, $d\alpha$ and $d\beta$ are elements of angles that define the contour lines of the surface element *dA* (see Figure 5). Therefore, using the approximation of small angles [$\sin(d\alpha/2) \cong d\alpha/2$] and based on the geometry depicted in Figure 5 one can easily derive that

$$p_l \cdot dA = p_{V.} \cdot dA + 2\gamma dl \cdot d\alpha/2 + 2\gamma dl \cdot d\beta/2. \quad (4)$$

The Young Laplace equation is obtained using the relationships between the angles $d\alpha$, and $d\beta$, the curvature radii, R_1 and R_2, and dl ($d\alpha = dl/R_1$, and $d\beta = dl/R_2$) in the equation (4).

$$p_l - p_V = \gamma \cdot \left(\frac{1}{R_1} + \frac{1}{R_2}\right). \quad (5)$$

The total curvature radius of the meniscus surface, R, is:

$$\frac{1}{R} = \frac{1}{R_1} + \frac{1}{R_2} \quad (6)$$

Thus, the Young-Laplace equation reads:

$$p_l - p_V = \frac{\gamma}{R}. \quad (7)$$

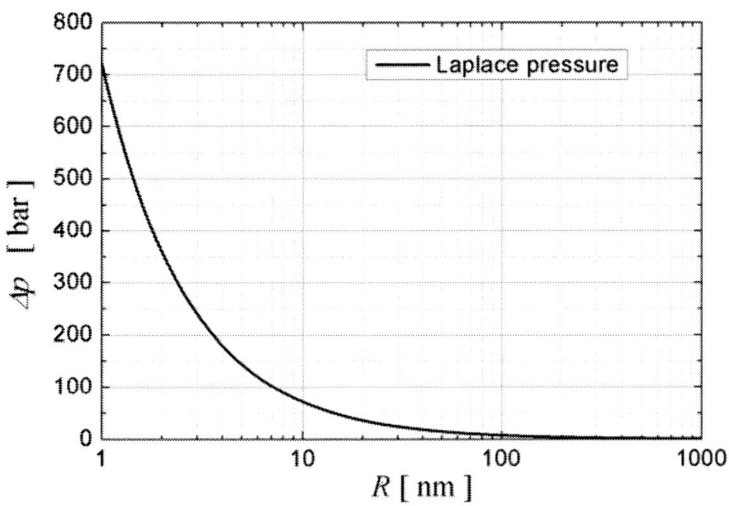

Figure 6. Dependence of the Laplace pressure, Δp, on meniscus curvature radius, R, for water at room temperature ($\gamma = 0.072$ N/m).

The total curvature radius of a surface depends on the curvature radii R_1 and R_2 in two perpendicular planes that cross along the line normal to the meniscus surface. R_1 and R_2 are negative for concave curvatures (bended towards vapor phase) and positive for convex curvatures (bended towards liquid phase).

Let us notice that for microscopically curved menisci the pressure difference, $\Delta p = p_l - p_v$, can have very large values. Thus, for a meniscus curvature radius of 10 nm, $\Delta p = 72$ bars. Figure 6 shows the dependence of Δp on curvature radius for convex menisci ($R > 0$ and $\Delta p > 0$). The same dependence is obtained for concave menisci if the signs of R and Δp are changed to minus.

3.2. Kelvin Equation

The Kelvin equation establishes the value of curvature radius of a liquid meniscus at thermodynamic equilibrium with vapor as a function of vapor pressure. Derivation of the equation is based on the thermodynamic equilibrium between thermodynamic potentials of liquid and vapor phases separated by the concave meniscus.

Let's express the change in the liquid free Gibbs energy ΔG_l per mol as result of meniscus bending from a flat to a curved surface as:

$$\Delta G_l = \int_{p_0}^{p} V_m \cdot dp = V_m \cdot \Delta p = V_m \cdot \frac{\gamma}{R}, \tag{8}$$

where V_m is the molar volume of the liquid, which is supposed to be constant during transformation. The change of Gibbs free energy per mole of vapor phase is:

$$\Delta G_V = \int_{p_0}^{p} V_m' \cdot dp = \int_{p_0}^{p} \frac{RT}{p} \cdot dp = RT \cdot ln\left(\frac{p}{p_0}\right), \tag{9}$$

where V_m' is the molar volume of vapor and R is the constant of the ideal gas (the vapor is considered as a perfect gas). At thermodynamic equilibrium the curvature radius of the meniscus is the Kelvin curvature radius ($R = R_K$) at which the two energy variations should be equal ($\Delta G_l = \Delta G_V$). This leads to

$$V_m \cdot \frac{\gamma}{R_K} = RT \cdot ln\left(\frac{p}{p_0}\right), \tag{10}$$

which is the Kelvin equation establishing a relationship between the meniscus curvature radius at thermodynamic equilibrium with vapor, R_K, and the relative pressure of the vapor, p/p_0. The equation is usually written as:

$$\frac{1}{R_K} = \frac{RT}{\gamma \cdot V_m} \cdot ln\left(\frac{p}{p_0}\right). \quad (11)$$

A practical formula for Kelvin radius of water menisci at room temperature (γ = 0.072 N/m, V_m = 18 ml/mol at T = 293 K) is:

$$R_K = \frac{0.54\ nm}{ln(p/p_0)}. \quad (12)$$

For water vapor in air, the relative pressure of water vapor expressed in percents is the relative humidity, RH, of air and, according to the Kelvin equation, its value determines the value of curvature radius of water menisci in thermodynamic equilibrium with vapor. Figure 7 presents a plot of Kelvin radius dependence on air humidity. One notices that R_k is negative for undersaturated water vapor ($p < p_0$) and its absolute value increases to infinity by the increase of RH towards 100%. The value of R_k is of the order of few nanometers at moderate values of air RH (R_k = 5.4 nm at RH = 90 %).

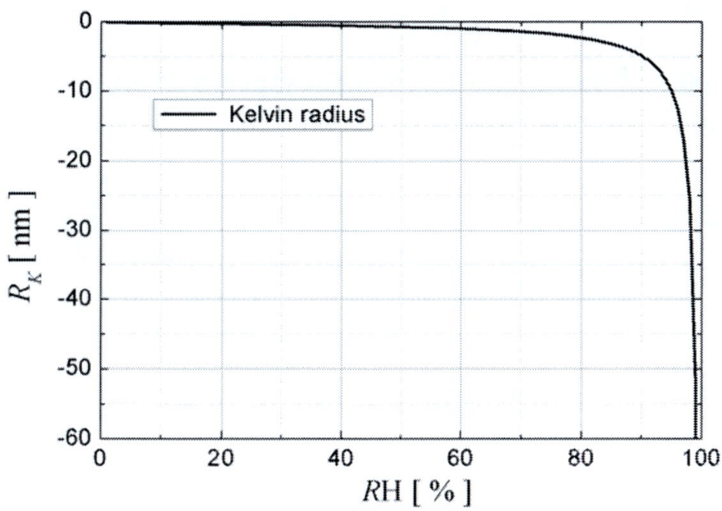

Figure 7. Dependence of Kelvin radius, R_K, on the relative humidity of air (RH = $100 \cdot p/p_0$ %), at room temperature.

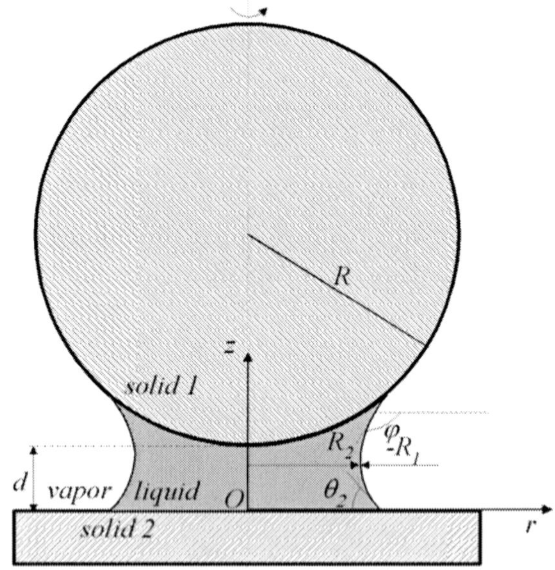

Figure 8. Geometry of a water bridge formed by capillary condensation in the nanoscopic gap between a spherical particle (solid 1) of radius R and a flat solid surface ($R = \infty$, solid 2). R_1 and R_2 are meniscus curvatures in vertical and horizontal planes, respectively, φ the meniscus surface tilting angle with respect to the horizontal plane, and z and r are cylindrical coordinates used for computing of meniscus shape.

3.3. Capillary Condensation at Single-Asperity Contact of Two Microscopic Solid Bodies

Very often capillary condensation of water at single-asperity contacts of two solid bodies is modeled for a small spherical particle in mechanical contact with a flat surface of a second body. This is happening to be the case of microscopic dust particles that adhere to the flat surface of a macroscopic body, but the model describes correctly also the water capillary condensation at the contact of the AFM tip (modeled by the spherical particle) and the sample (modeled by the flat surface). In short time after contact, the water vapor from the humid air condenses and builds up a water neck that surrounds the contact of the two bodies (Figure 8). The system has a cylindrical symmetry and the water meniscus has a concave curvature in vertical plane ($R_1 < 0$) and convex curvature in horizontal plane ($R_2 > 0$).

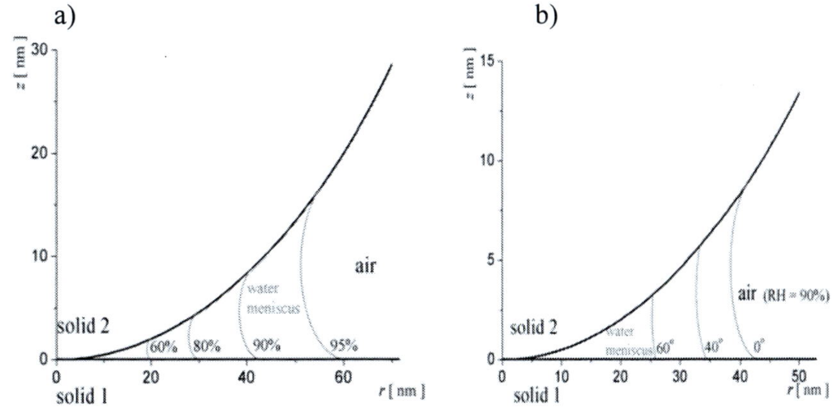

Figure 9. a) Grow of a water meniscus at the nanocontact of two wetting solid bodies in air by the increase of RH. b) Dependence of the water meniscus on wetting properties of the solid bodies in contact (RH = 90% and water contact angle of both solid bodies were taken 0°, 40° and 60°). The water menisci were computed by integral of Laplace equation for values of R_K corresponding to RH values.

At thermodynamic equilibrium the total curvature radius of the water meniscus is determined by Kelvin equation and should be the same on every point on the meniscus surface. The value of meniscus curvature determines the amount of water condensed at the contact. However, the size of the water meniscus is determined also by the wetting property of the two solid bodies in contact, i. e. by the values of water contact angle of the two solid bodies in contact. The amount of the condensed water increases with the increase of the air humidity and decrease of water contact angle values of the two bodies. Computation of the amount of the capillary condensed water requires finding of a solution of Young-Laplace equation for a rotationally symmetric meniscus with total curvature radius R_K and border conditions satisfying the value of water contact angles of the two solids. In this case the differential equations of the meniscus shape, *z(r)*, in vertical plane is [17]

$$\begin{cases} \dfrac{dr}{d\varphi} = \dfrac{\cos(\varphi)}{1/R_k - \sin(\varphi)/r} \\ \dfrac{dz}{d\varphi} = \dfrac{-\sin(\varphi)}{1/R_k - \sin(\varphi)/r} \end{cases} \tag{13}$$

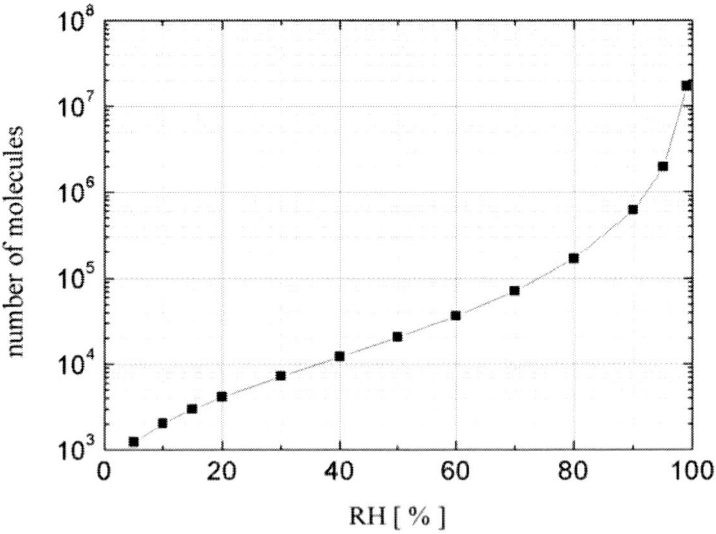

Figure 10. Grow of the water meniscus size at the contact of a spherical particle (R = 100 nm) with a flat surface by the increase of air RH. The solid body surfaces taken in computation of the meniscus Laplace equation were considered perfectly hydrophilic (water contact angle was assumed to be zero).

The integration variable, φ, is the tilting angle of meniscus surface with respect to the horizontal plane. The equations (13) were numerically integrated using a forth-order Runge-Kuta routine with fixed step size and the obtained solutions for perfectly wetting solid bodies (zero water contact angle values) at different values of RH and for RH = 90° at different water contact angle values of the two bodies are presented in Figure 9 a) and b), respectively.

The size of the meniscus increases by the increase of RH and decreases by the decrease of water contact angle of the two solid bodies. A variation of the number of condensed water molecules with RH at the nanocontact of a perfectly wetting particle with radius of 100 nm with a perfectly wetting flat surface is presented in Figure 10. The plot shows that the number of water molecules that condense at the nanoscopic contact is large enough to justify the use thermodynamic equations and methods of statistical physics in study of capillary condensation phenomenon. It is noticed that, except very large value of RH, the number of condensed water molecules increases by a decade at each increase of RH by about 30%.

4. CAPILLARY FORCE

Formation of the capillary water bridge at the single-asperity nanocontact of two solid bodies generates an attractive (adhesion) force called the capillary force. To understand the origin of this force and to find its mathematical expression let us consider the model of the capillary water bridge that surrounds the contact between a spherical microscopic particle and a flat surface (Figure 11) and recall the fact that according to the Young-Laplace equation the pressure inside the concave water meniscus is smaller than the vapor pressure with $\Delta p = \gamma/R$, where R is total curvature of the water meniscus, which at thermodynamic equilibrium is equal to the Kelvin radius. It is this pressure difference that generates the capillary force. If R_2 ($R_2 << R$) is the radius of the meniscus contact line with the spherical particle and we consider that Δp acts uniformly on a surface πR_2^2, than the capillary force is:

$$F_c = \pi R_2^2 \cdot \Delta p = \frac{\pi \gamma R_2^2}{R_K} \tag{14}$$

where R_k is the total curvature of the meniscus. The principal curvature radii of the meniscus, $-R_1$ in vertical plane and R_2 in horizontal plane, at thermodynamic equilibrium satisfy the equation:

$$\frac{1}{R_1} + \frac{1}{R_2} = \frac{1}{R_K} \tag{15}$$

Generally, R_1 and R_2 have different values for different positions on the meniscus surface. However, for water menisci much smaller than the solid bodies in contact ($R >> |R_2| >> |R_1|$), the capillary force can be roughly estimated by the so called geometrical approximation of the meniscus shape. In this approximation the meniscus shape in vertical plane is considered as a portion of a circle with radius $-R_1$. Variation of R_2 value on the meniscus surface is neglected and R_2 is considered as the radius of the meniscus contact line with the solid bodies. Moreover, since $|R_2| >> |R_1|$, $R_1 \cong R_K$. With this approximation, is easy to show that:

$$R_2^2 \cong 2R \cdot R_1(\cos\theta_1 + \cos\theta_2), \tag{16}$$

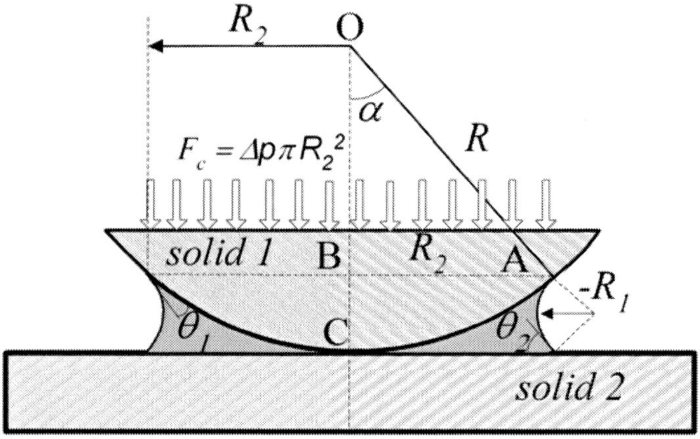

Figure 11. Capillary force generated by Laplace pressure, $\Delta p = p_V - p_l$, of a water meniscus formed at the contact of a spherical particle with a flat surface.

where θ_1 and θ_2 are the water contact angles of the two solid bodies. The equation (16) is derived from Pythagoras's theorem applied in the triangle OBA (see Figure 11) where the length of |BC| is estimated as $R_1(\cos\theta_1 + cos\theta_2)$. The finite size of the angle α is neglected ($\alpha \cong 0$). Using the equation (16) and the approximation $R_l \cong R_K$ in the equation (14) derives the formula of O'Brien and Hermann [18] of capillary force:

$$F_c = 2\pi\gamma \cdot R(cos\theta_1 + cos\theta_2). \tag{17}$$

If $\theta_1 = \theta_2 = \theta$, the equation (17) reads

$$F_c = 4\pi\gamma R cos\theta \tag{18}$$

The equation shows that the capillary force is proportional to the size of particles in contact (R can be considered the equivalent radius of two spherical particles in contact or, in case of AFM measurements, the equivalent radius of the AFM tip and sample: $1/R = 1/R_t + 1/R_s$, where R_t is the AFM tip radius and R_s is the local curvature of the AFM sample). The equation shows also that F_c is proportional with the superficial tension of water and depends on hydrophilicity of the solid surfaces in contact. No capillary meniscus and no capillary force exist between particles with non-wetting surfaces ($\theta \geq 90°$). However, the formula is not accurate and a more careful analysis should take in account also the superficial tension force acting at the contour lines of the meniscus with the solid bodies. The superficial tension force acts in the plane of the meniscus surface at the contour line with the one of the solid bodies in contact. Only the components of the superficial tension forces acting along the vertical direction contribute to the capillary force (Figure 12). Thus, the superficial tension force acting at the contour line of the water meniscus with the spherical particle is:

$$F_{\text{sup}} = 2\pi \cdot R_2 \cdot \gamma \sin(\alpha + \theta_1) \tag{19}$$

The capillary force is the sum of the Laplace force and superficial tension force. Thus, a more accurate formula of the capillary force [19] is:

$$F_a = -\frac{\pi \cdot R_2^2 \cdot \gamma}{R_K} + 2\pi \cdot R_2 \cdot \gamma \sin(\alpha + \theta_1) \tag{20}$$

However, in order to compute the capillary force a value of the radius of the meniscus contact line with the solid bodies in contact is required. This can be approximately computed by the equation (16) or can be found from the solution of the Young-Laplace differential equation of the meniscus [20].

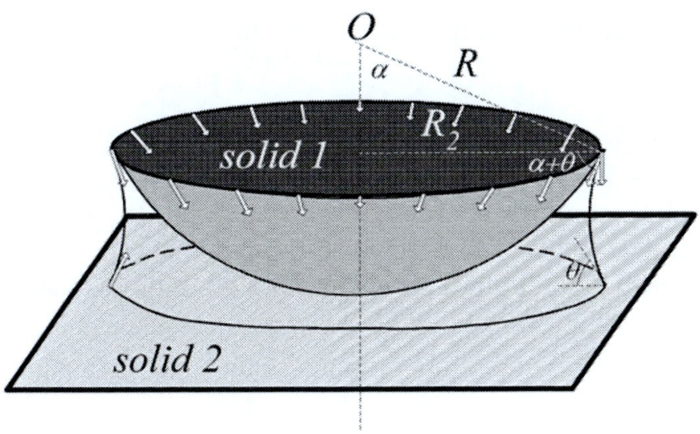

Figure 12. Superficial tension force acting along the contour line of the meniscus with the spherical particle. Note that the superficial tension force acts also at the contour of the meniscus with the flat surface. The superficial tension forces acting on the two contour lines should be equal and opposite (according to the third principle of the Newtonian mechanics).

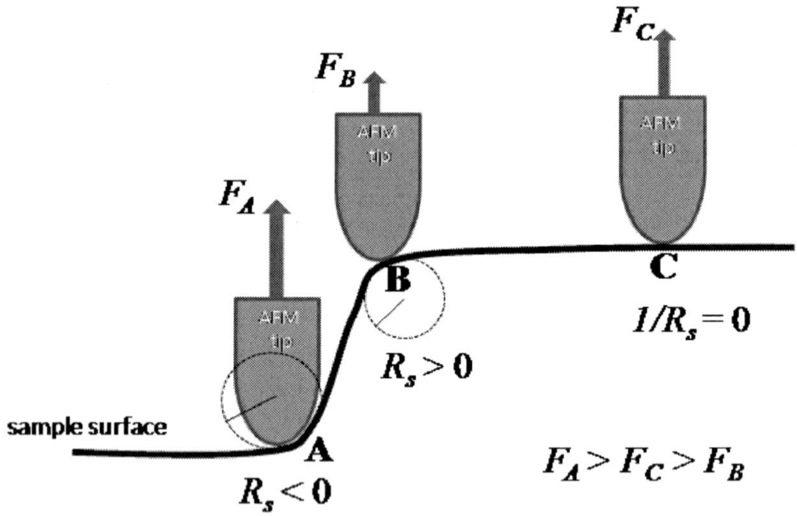

Figure 13. Dependence of the capillary force on sample topography. The sample local curvature radius changes from the bottom edge of a pit (A), where $R_s < 0$ to the top edge of the pit (B), where $R_s > 0$, to a flat surface (C), where $R_s = \infty$. As a result, the pull-off force required to detach the tip from the sample surface is larger in A than in C or B.

4.1. Dependence of the Capillary Force on Contact Curvature Radius

One of the main characteristics of the capillary force described by the approximate O'Brien and Hermann equation is the force proportionality with the curvature radius of the solid bodies in contact. For the case of the AFM measurements the total curvature radius of the tip and sample surfaces should be taken into account [19]. The total curvature radius of the tip and sample, R, is:

$$- \quad - \quad - \qquad (21)$$

where R_t and R_s are the tip and sample, respectively, curvature radii. While R_t is always positive (the surface of the AFM tip is convex) and constant in a certain AFM experiment, the sample local curvature radius change from one point of the sample to other and can take also negative values (for the concave local curvature of the sample). The dependence of the capillary force on R_s was illustrated experimentally by the analysis of capillary force distributions on a standard grating sample (silicon covered by platinum) with rectangular pits [19]. The capillary force was much larger for the AFM tip in contact with the concave surface at the bottom of the pit edge (Figure 13 A) than one measured on the flat surface (Figure 13 C), which at its turn was larger than one measured on the convex surface of the sample surface at the top of the pit edge (Figure 13 B).

4.2. Dependence of the Capillary Force on Air Humidity

According to the approximate equation (18), the capillary force does not depend on air humidity. This result is somehow counterintuitive, because as it is shown in the precedent section, the size of the water meniscus increases steeply by the increase of RH. However the increase of the meniscus size with the increase of RH is compensated by the corresponding decrease of the Laplace pressure (which decreases with the increase of RH). The results of numerical computation of the capillary force based on the solutions of Young-Laplace equation for rotationally symmetric menisci with values of the total curvature radius R_K corresponding to different values of RH shows a slight decrease of the capillary force by the increase of RH (Figure 14). To illustrate the contribution of the superficial tension force to the total capillary force (closed symbols), the Laplace component (open symbols) of the capillary force was also plotted in Figure 14. Moreover, the computed forces are compared with the approximate capillary force equation (18). The data shows that the contribution of the superficial tension force to the capillary force is relatively small and this contribution increases with the increase of RH and decrease of surface hydrophilicity.

The dependence of capillary force on RH was experimentally checked by AFM measurement of pull-off force between an AFM probe (NSC 36 from Mikromash, Inc.) and a flat sample [silicon wafer Si(1,0,0)]. The measurements were performed by a commercial AFM apparatus with an environmental chamber. Relative humidity (RH) of the measurement medium was controlled by flowing high purity nitrogen and water vapor through the environmental chamber. The obtained experimental results are presented in Figure 15 and confirm the theoretical dependence of capillary force on RH.

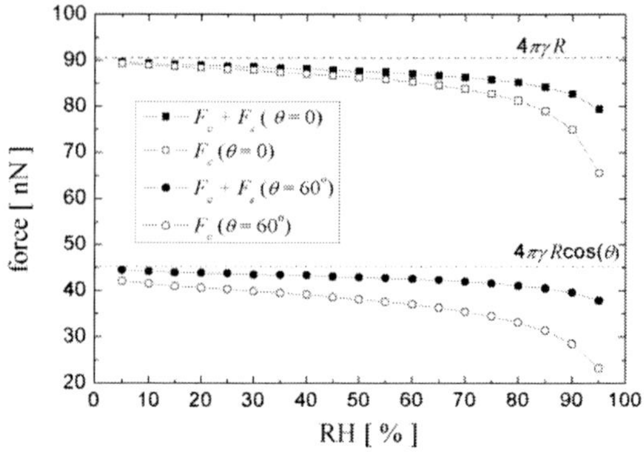

Figure 14. Dependence of the capillary force on air relative humidity as computed by numerical solution of Young-Laplace equation of water menisci in thermodynamic equilibrium with water vapor at different RH.

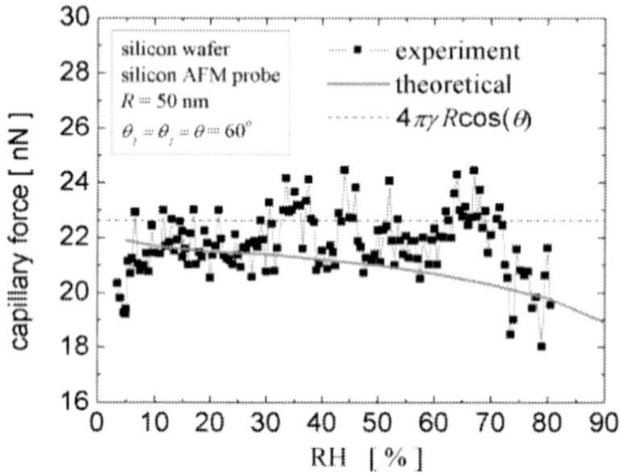

Figure 15. Comparison of theoretical and experimental values of capillary force between an AFM tip (silicon tip with curvature radius of 50 nm and water contact angle of 60°) and a sample (polished silicon wafer).

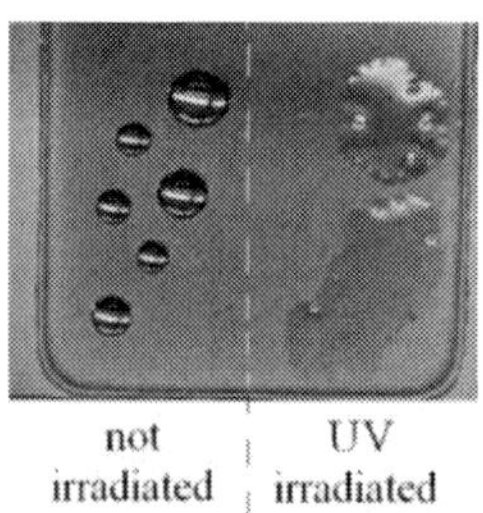

Figure 16. Effect of UV irradiation on the hydrophilicity of a titanium dioxide thin film surface.

4.3. Dependence of the Capillary Force on Surface Hydrophilicity

As it is described in the precedent section, the size of the water meniscus formed by capillary condensation at the microscopic contact of two bodies increases by the increase of body hydrophilicity. This dependence was experimentally checked by an AFM study of the capillary force at constant RH between titanium dioxide (TiO_2) covered AFM probes and samples. Titanium dioxide is a material known to improve its hydrophilicity by UV radiation [21]. Figure 16 shows a photo of sessile water drops on UV irradiated and non irradiated parts of a titanium dioxide thin film which proves that the hydrophilicity of the titanium dioxide thin films improved very much as result of UV irradiation.

Commercial AFM probes (Olympus, OMCL-TR800PSA-1) and silicon wafers were covered by ultra thin films of amorphous titanium dioxide (TiO_2). The samples consisted of ultra thin films of amorphous TiO_2 deposited on silicon wafers [Si(100)] by radio frequency magnetron sputtering deposition in argon-oxygen gas at a pressure value of 10 mTorr at a discharge power of 50 W. Details on the deposition process, film characterization and film UV light induced hydrophilicity were published elsewhere[22]. The hydrophilicity of the AFM tip and sample surfaces was characterized by measurements of the contact angle of sessile drops of pure distilled water (pH = 5.8) deposited on the base of the AFM probe and sample surfaces. As result of UV light irradiation during 20 minutes, the water contact angle decreased from 60^0 to about 40^0 for either AFM probe or sample surfaces. Measurements of cantilever resonant frequency before and after deposition showed that there was no noticeable change of the force constant of the cantilever (0.57 N/m) as a result of TiO_2 deposition.

The capillary force was measured by pull-off experiments conducted at different values of water contact angles of the AFM tip and sample surfaces in atmospheric air at constant temperature (26°C) and relative humidity (RH = 32 ± 2%). Figure 17 shows the histograms of the pull-off force values determined for 100 points randomly placed on the sample surface. The histogram plots show that the adhesive force increased by the increase of the hydrophilicity of the AFM tip and sample surfaces. A dispersion of the pull-off force values that increased by the increase of the hydrophilicity of the interaction surfaces is also noticed. The larger adhesive force values measured for the surfaces with increased hydrophilicity may be attributed to the increase of the size of the water neck formed by capillary condensation at the tip-sample contact. The wide dispersion of the tip-sample adhesive force values for the surfaces with increased hydrophilicity may be related to the properties of the growing film of absorbed water molecules on surfaces. Because of the possible heterogeneity of the water film adsorbed on the surfaces with increased hydrophilicity, water menisci of a relatively large variety of shapes and sizes may form at the contact of these surfaces with the effect of widening of the pull-off force value histograms.

The experimental data was compared with the theoretical dependence of capillary force derived from the use of the expression (20) of the capillary force and solutions of the differential equation (13) of the meniscus shape for different values of the water contact angle of the sample. This comparison is shown on Figure 18. As expected, the numerical computation of the capillary force confirms the experimental data and shows a decrease of the capillary force by the increase of the water contact angle of the sample.

The dependence of capillary force on sample surface hydrophilicity is very important for the AFM characterization of chemical contrast of sample surfaces by measurements of adhesive and friction force distributions on sample surfaces. The friction force experienced in AFM scanning of sample surfaces is in certain conditions proportional to the adhesive force, so that both forces can be used to determine contrast between hydrophilic and less hydrophilic microscopic regions on a sample surface. A good contrast in tip-sample adhesion or friction forces reveals chemical differences between regions with different hydrophilicity. This effect was used for microscopic characterization of titanium dioxide thin film hydrophilicity [20] or for characterization of chemical nanopatterns obtained by nanolithography [23].

An increase of the hydrophilicity of the AFM silicon probes is easily obtained by cleaning their surfaces in plasma, piranha solution, or by UV irradiation [24]. All these methods oxidize and remove the airborne hydrocarbon contaminants absorbed on the AFM probe surfaces and render clean and very hydrophilic surfaces. Figure 19 shows the effect of AFM probe cleaning on the capillary adhesive force measured by pull-off experiments in air (RH = 30%) with a commercial AFM probe (PPP-BSI from Nanosensors^{TM}) with sharpened tip (nominal radius of 5nm) on glass. As result of plasma cleaning, the contact angle of the AFM probe surface (silicon) decreased from about 90° to about 10°. This resulted in an increase of the capillary force with about 50%. In this particular experiment the adhesive force values are small because of the very small value of the AFM tip radius.

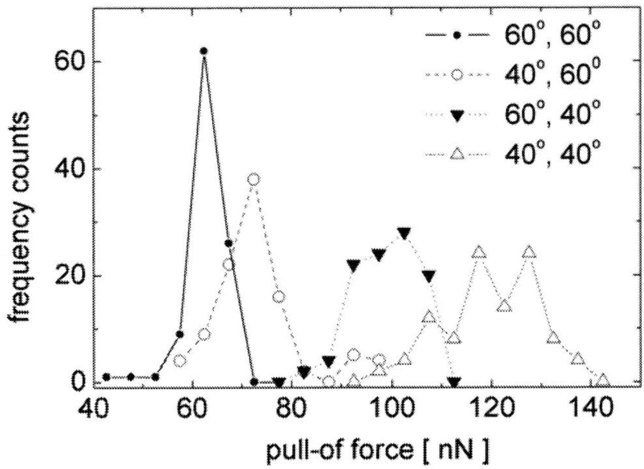

Figure 17. Hystograms of capillary force values measured by AFM measurements of the pull-off force between titanium dioxide covered AFM tip and sample surfaces at different values of water contact angle. The water contact angle values for the AFM tip are shown on left column while the values for the sample surface are shown on right column.

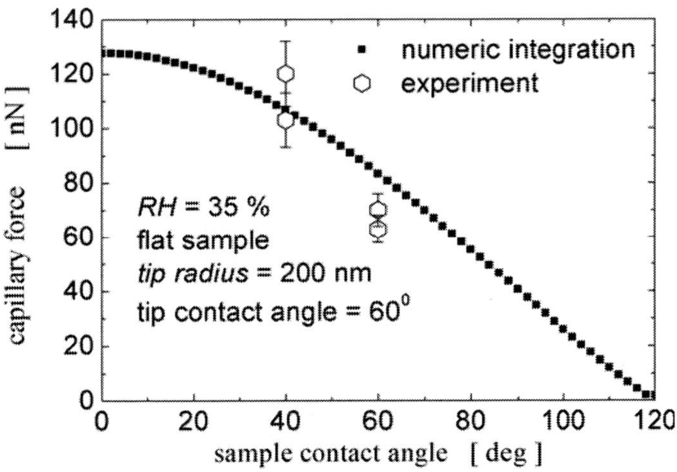

Figure 18. Dependence of the capillary force on the water contact angle of the sample as was determined by numerical computation of the water meniscus. Experimental data obtained for two values of the water contact angle of the sample or AFM tip are shown for comparison.

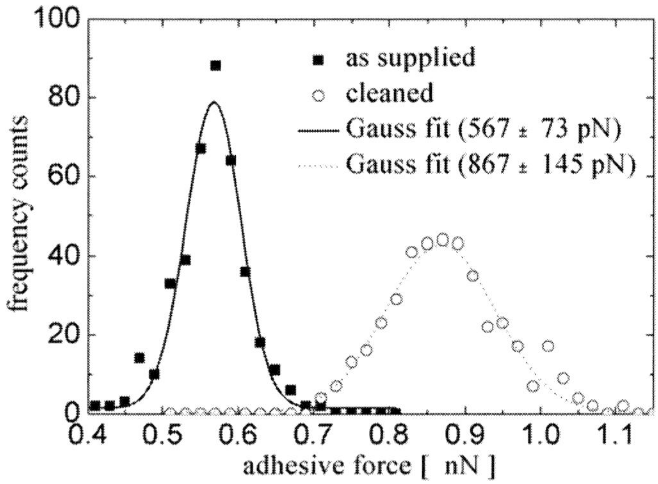

Figure 19. Histograms of adhesive force measurements performed with as supplied commercial probe (PPP-BSI from Nanosensors™) and the same probe after plasma cleaning on a glass sample.

4.4. Dependence of the Capillary Force on the Tip Sample Distance: Probing the Water Meniscus Volume

When the AFM tip is retracted the tip-sample distance increases and the water meniscus is thinning and at certain distance, called meniscus break up distance, breaks up. We have performed an experimental and theoretical study of thinning and breaking of water menisci during rupture of nanoscale contacts [25]. The study, which was based on analysis of force-distance curves obtained in AFM pull-off experiments, concluded that a nanoscale contact rupture causes a stretch of the water meniscus, with a negligible evaporation, up to a distance that is much larger than $2 \cdot R_K$, which is breakup distance of a meniscus in thermodynamic contact with the water vapor. Based on this experimental observation, a theoretical model for the water meniscus elongation was proposed. The model considered a set of numerical solutions of the meniscus Young-Laplace equation that self-consistently assumed for the increase of the tip-sample separation distance a decrease of the meniscus curvature, a constant meniscus volume and retraction of the contact lines at the tip and sample surfaces. Figure 20 present a set of such numerical solutions for the water meniscus shape during the AFM tip retraction.

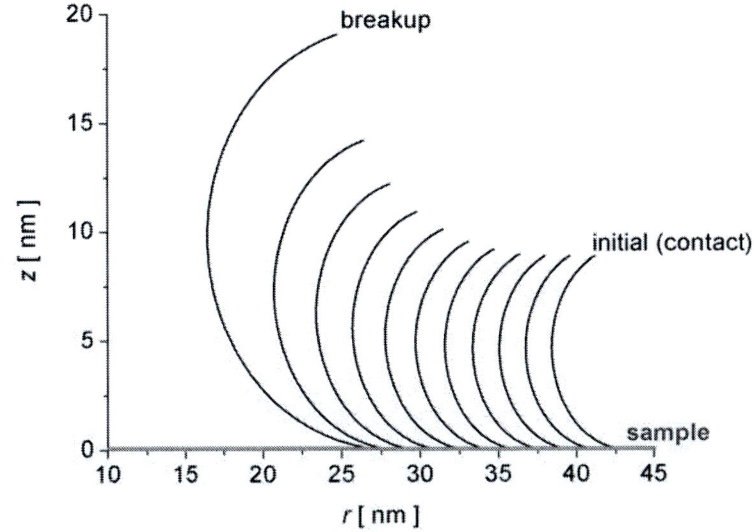

Figure 20. Computed shape of water meniscus at the AFM tip retraction from contact until the breakup distance (R_t = 50 nm, RH = 90%, $\theta_t = \theta_s = 20°$).

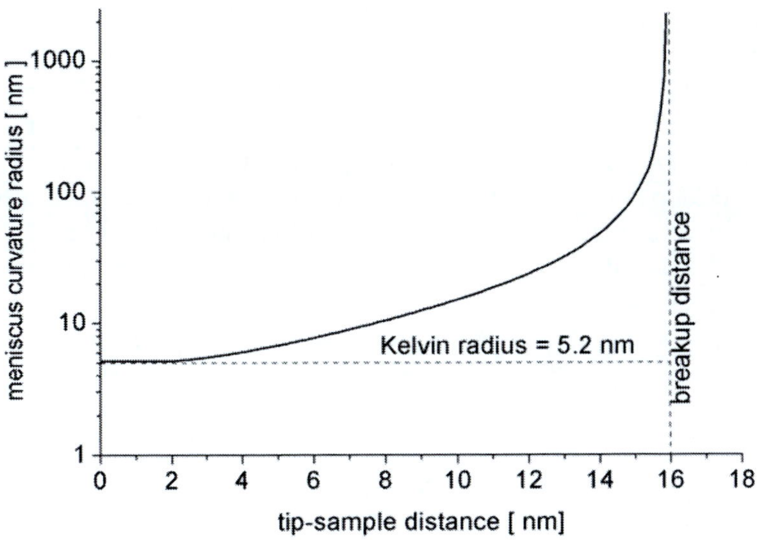

Figure 21. Change of the meniscus curvature during stretching (R_t = 50 nm, RH = 90%, $\theta_t = \theta_s = 20°$).

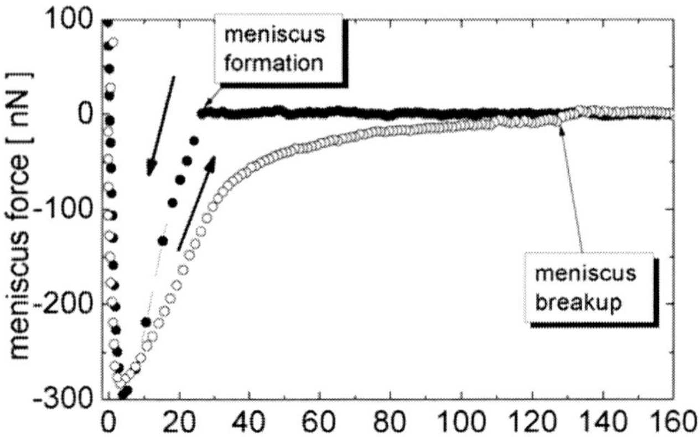

Figure 22. Experimental evidence of monotonic decrease of the capillary force during meniscus stretching (R_t = 200 nm, RH = 90%, silicon tip, freshly cleaved mica sample).

Figure 21 shows how meniscus curvature changes during stretching until the break up distance. Decrease in meniscus curvature radius is accompanied by a decrease of the Laplace pressure and thus of the capillary force. During stretching, the meniscus was considered not in thermodynamic equilibrium with the water vapor, but in mechanical equilibrium with the meniscus external force, which means that the external force that acts to break up the contact counterbalances the capillary and superficial forces of the meniscus. The model predicts a monotonic decrease towards zero of the meniscus capillary force when the tip-sample separation distance is increased towards the meniscus breakup distance, d_b.

The AFM pull-off experiments showed that the capillary force decreased monotonically to zero in agreement with the theoretical prediction as is shown in Figure 22. Results of numerical computation of water meniscus breakup showed that the breakup distance is roughly equal to the cubic root of the water meniscus volume, which is a result that was known to be valid for stretching of nonvolatile liquid droplets.

Conclusion

Atomic force microscopy is one of the main techniques used in study of the capillary condensation of water at nanoscopic contacts. The phenomenon occurs in short time after the very sharp tip of the AFM is brought in contact with the sample surface and results in formation of a water meniscus that surrounds the contact. The size of the water meniscus depends on air humidity, which determines the curvature radius of the meniscus in thermodynamic equilibrium with the water vapor. Increase of the air humidity results in increase of the water meniscus size. The most evident effect of capillary condensation at AFM tip-sample contact is the capillary force. The capillary force is generated by the water neck that surrounds the tip-sample contact mainly as an effect of the Laplace pressure of the meniscus. The capillary force is directly measured by AFM pull-off experiments. In this type of experiments, the AFM tip is approached to the sample surface until contact. Then, the movement of the AFM probe is reversed and the AFM tip is retracted until the initial position. The maximum attractive force experienced during tip retraction is considered as the tip-sample capillary force. Pull-off experiments proved that capillary force decreases slightly with the increase of air relative humidity, is proportional with the tip-sample curvature radius and depends greatly on hydrophilic properties of AFM tip and sample surfaces. This last property of the capillary force was used for study of surface hydrophilicity at nanoscale. Breakup of the AFM tip-sample contact during the AFM tip retraction cause stretching of water menisci until a certain distance that experimentally was proven to be much larger than twice of the Kelvin radius. This means that the meniscus curvature changes during meniscus stretching, which means that the stretched meniscus is not in thermodynamic equilibrium with water vapor. Analysis of capillary force dependence on stretching distance gives a clue on the value of water meniscus volume.

REFERENCES

[1] W. T. Thomson, *Phil. Mag.* 42, 448 (1871).

[2] Bhushan, Bharat *Wear* 1999, 225–229, 465–492.

[3] Schenge, M.; Li, X.; Schaefer, J. A. *Tribology Lett.* 1999, 6, 215-220.

[4] de Lazzer, A.; Dreyer, M.; Rath, H. J. *Langmuir* 1999, 15, 4551-4559.

[5] Piner, Richard D.; Zhu, Jin; Xu, Feng; Hong, Seunghun; Mirkin, Chad A. *Science* 1999, 283, 661-664.

[6] Shao, Z. F.; Mou, J.; Czajkowsky, D. M.; Yang, J.; Yuan, J. Y. *Adv. Phys.* 1996, 45, 1-86.

[7] Blackman, G. S; Mate, C. M.; Phylpott, M. R. Phys. Rev. Lett. 1990, 65, 2270

[8] Grigg, D. A.; Russell, P. E.; Griffith, J. E. *J. Vac. Scie. & Tech. A: Vacuum, Surfaces, and Films* 1992, 10, 680.

[9] G. Binnig, C.F. Quate, C. Gerber, *Phys. Rev. Lett.* 56 (1986) 930.

[10] M. Schenk, M. Futing, R. Reichelt, *J. Appl. Phys.* 1998, 84, 4880.

[11] L. Xu, A. Lio, J. Hu, D. Ogletree, M. Salmeron, *J. Phys. Chem. B* 1998, 102, 540.

[12] S. Biggs, R. G. Cain, R. R. Dagastine, N. W. Page, *J. Adhesion Sci. Technol.* 16 2002, 869.

[13] M. J. Adams, B. J. Briscoe, J. Y. C. Law, P. F. Luckham, D. R. Williams, Langmuir 2001, 17, 6953.

[14] Burnham, N. A.; Colton, R. J.; Pollock, H. M.; *Nanotechnology* 1993, 4, 64.

[15] Sounilhac, S.; Barthel, E.; Creuzet, F. *J. Appl. Phys.* 1999, 85, 222.

[16] Hans-Jurgen Butt, Karlheinz Graf, Michael Kappl, *Physics and Chemistry of Interfaces*, Wiley-VCH Verlag GmbH &Co. Weinheim 2003.

[17] Hartland, S; Hartley, R.W. *Axisymetric fluid-liquid interfaces*; Elsevier: Amsterdam, 1976.

[18] W.J. O'Brien, J.J. Hermann, *J. Adhes.* 5 (1973) 91.

[19] L. Sirghi Langmuir 1.

[20] L. Sirghi, M. Nakamura, Y. Hatanaka, O. Takai, *Langmuir* 17 (2001) 8199-8203.

[21] Fujishima, A. and Zhang, X., 2006, *Comptes Rendus Chimie* 9 (5-6), 750-760.

[22] L. Sirghi; T. Aoki, Y. Hatanaka, Thin Solid Films 422 (2002) 55-61.

[23] L. Sirghi, F. Bretagnol, S. Mornet, T. Sasaki, D. Gilliland, P. Colpo, F. Rossi *Ultramicroscopy* 109 (2009) 222–229.

[24] L. Sirghi, O. Kylian, D. Gilliland, G. Ceccone, F. Rossi, *J. Phys. Chem.* B 110 (2006) 25975-25981.

[25] L. Sirghi, R. Szoszkiewich, E. Riedo, *Langmuir* 22 (2006) 1093.

In: Water Condensation
Editor: Amimul Ahsan
ISBN: 978-1-61324-874-4

Chapter 3

RECENT ADVANCES IN THE SYNTHESIS OF HETEROCYCLIC COMPOUNDS IN AQUEOUS MEDIUM

Y. V. D. Nageswar**, K. Harsha Vardhan Reddy, S. Narayana Murthy and A. Ashwan Kumar
Organic Chemistry Division-I, Indian Institute of Chemical Technology, Hyderabad, India

ABSTRACT

The widespread application of structurally diverse heterocycles in drug development has led to the exploration of new synthetic strategies. Demand for efficient methods for novel heterocyclic systems has become a powerful driving force for advances in synthetic organic chemistry. In view of the significance attached to green chemistry and its relevance to the present day problem of global warming, the development of novel, simple, cleaner synthetic protocols is attracting attention in both academic and industrial research, resulting in an ever increasing number of publications or reports on this topic. Designing environ friendly synthetic strategies in water, minimizing the use of harmful, toxic, and flammable organic solvents and hazardous reagents/catalysts, is attaining the priority over other issues. Even though water has the status of universally acceptable solvent, its utility as a solvent medium for organic transformations has severe limitations, due to the solubility problems. The present account is an attempt is to revisit briefly into several recent aspects of water medium reactions involving the synthesis of potential heterocyclic systems, which will be useful to research community.

*E-mail: dryvdnageswar@gmail.com.

INTRODUCTION

In general organic synthetic processes are conducted in a solvent medium, which is of toxic, inflammable and hazardous nature. Experimental procedures invariably involve spillages, evaporation as well as recycling resulting in environmental pollution, which has become a major issue to be addressed at global level. These solvents may be chlorinated hydrocarbons or polar aprotic solvents such as dimethylformamide or dimethyl sulfoxide etc., Apart from the atmospheric pollution, use of solvent media in an industrial process also causes health hazards directly affecting the people around. Another important criterion to be considered in this aspect is the groundwater contamination, being caused by the solvents, wherever they are used, in the form of effluents.

In view of several such issues involved in the usage of a wide range of volatile, non volatile, inflammable, nonpolar as well as polar aprotic solvents, design and development of environmentally benign sustainable reaction processes, involving non-conventional reaction media is widely accepted to become a global necessity. This paradigm shift from classical organic reaction processes to ecofriendly technology led to the evolution of green chemistry. In this context, apart from other parallel issues concerning reaction processes, the main subject under consideration is the selection of alternate solvent, which can effectively be implemented in an organic reaction.

Water is the nature's solvent and all the biological syntheses are being carried in aqueous media. It is an essential element of life on our planet. Water possesses unique structural form, which bestows distinguished physical and chemical properties. It exhibits powerful hydrogen bonding, large dielectric constant, high heat capacity as well as a broad temperature range to remain in liquid state. However, in spite of several such advantages, due to the complex nature of present day organic syntheses, being conducted in academic as well as industrial organizations, water failed to become solvent of choice. Most of the present day organic reactions could not be conducted in water due to the hydrophobic nature of organic molecules, as solubility is the prerequisite for the reactivity. Wohler's urea synthesis as well as Baeyer and Drewsen's indigo synthesis represent earliest examples of aqueous medium organic reactions. [1] Breslow's breakthrough investigation on Diels-Alder reactions in water medium has attracted the attention of researchers across the globe. Even though water is reported to enhance and accelerate the reaction rates and also increase the selectivity in a wide range of organic reactions, [2] it is not yet widely used as sole solvent in organic synthesis.

However several reactions are reported employing organic co-solvents, to increase the solubility of the organic compounds in water. [3] Introduction of polar functional groups on reactant molecules to enhance water solubility is another manipulation. [4] Moreover, aqueous organic reactions can also be conducted in either heterogeneous or homogenous phase in which volume of water may differ from moderate to large quantity. Apart from the above techniques, employing different additives like surfactants such as Triton-X, Aliquat-336 and PTS is also in vogue. [5]

In view of the advantages derived by the utilization of water as solvent medium, there is an increasing trend in performing organic synthesis in water. Apart from being ecofriendly, water also has the potentiality to become universally acceptable solvent, due to its abundance, and low cost. In recent years certain nitrogen containing hetero cyclic molecules are synthesized employing water as a preferred solvent. [6] It is of considerable interest to learn more about varied green technology platforms, which have been utilized for the synthesis of important hetero cyclic scaffolds. Being an important and growing area of research, aqueous organic synthesis has been regularly reviewed. [7] The present attempt is to review briefly recent water mediated organic reactions, resulting in the synthesis of various hetero cyclic frame works. Moreover to highlight the significance as well as utility of water as a sole solvent, this article limits itself to the publications, where in another additive or co-solvent has not been used.

FIVE MEMBERED HETEROCYCLIC SYSTEMS

A. Pyrazole Derivatives

Pyrazolone structural motif is an important constituent occurring in a wide range of pharmaceutical compounds. Among the substituted pyrazolones, heterocycle substitution increases protein kinase inhibitory activity resulting in the potentially active drug candidate molecules, which can be of use in cancers, proliferative diseases, psoriasis as well as arthritis. [8] Chromene is another significant structural moiety particularly present in many medicinally active compounds exhibiting varied types of properties including pro-apoptotic activity against cancer cells, [9] apart from spasmolytic, diuretic, antianaphylactic, as well as insecticidal activity. [10] In a recent research report, Gnanasambandam Vasuki et al described [11] a very efficient, facile, one pot approach for the synthesis of 4-pyrazolyl-4H-chromenes(I). In this important catalyst free, four component synthetic strategy, which can be seen as useful contribution for green chemistry, 2-hydroxy benzaldehyde, malononitrile, ethylacetoacetate, hydrazine hydrate were reacted in water as sole solvent medium at 30 °C, affording novel 2-amino-4-(5-hydroxy-3-methyl-1H-pyrazole-4-yl)-4H-chromene-3-carbonitrile derivatives in appreciable yields, where in pure products are isolated by simple filtration technique.

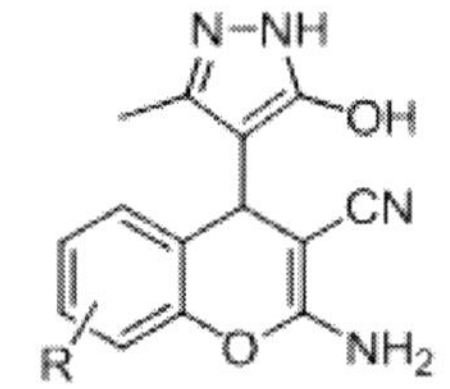

R = CH_3; OCH_3; C_2H_5; Cl; Br; OH; NO_2;

(I)

Pyrano pyrazoles acquired prominence due to their applications as pharmaceutical ingredients, and agrochemicals. [12] In a recent work by Gnanasambandam etal described [13] an efficient, ecofriendly, four component protocol in aqueous medium for the synthesis of a series of 6-amino- 5-cyano-3-methyl- 4-aryl/heteroaryl-2H,4H- dihydro pyrano[2,3-c] pyrazoles(II). This single step approach involves the reaction of hydrazine hydrate, ethylacetoacetate, aromatic aldehyde and malononitrile and the cyclized pure product is obtained simply by recrystallization technique, without any complicated chromatographic methods. Scope of the investigation was extended to observe the influence of different bases as well as solvents on the progress of the reaction. It was noted that the condensation resulted in excellent yields, when conducted in water as sole solvent medium at room temperature. Several aldehydes were used while fixing the parameters and the protocol is compatible with aliphatic and heterocyclic carbaldehydes also. Initially 1, 3 diketone reacts with hydrazine hydrate to result in the pyrazolone derivative, which undergoes Michael addition on the intermediate Knovenagel product, formed by the interaction between aldehyde and malononitrile. Cyclization and tautomerization of the intermediate finally results in the pyranopyrazole derivatives. This simple attractive one pot approach offers flexibility to build up a spectrum of highly substituted molecules of importance, in an ecofriendly way.

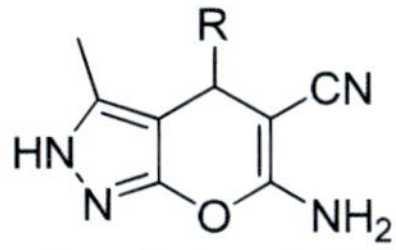

R = Ph; Substituted phenyl; 3-Pyridinyl; 4-Pyrimidyl; 2-Furanyl; 2-Thiophenyl; Isopropyl

(II)

It is well recognized that spiroindoline moiety highly enhances the biological activity [14] and the spiroindole skeleton exists as core structure in many natural products as well as pharmacological agents [15]. Spiroheterocycles with indole as well as pyran skeletal system are well recognised due to their analgesic, anti bacterial, anti convulsant and herbicidal properties. [16] A. Bajgir et al in an interesting report described [17] the synthesis of spiro[indoline-3,4'-pyrano[2,3-c]pyrazole]-5'-carbonitriles(III) by a one pot four component reaction in which β-ketoesters, hydrazine hydrate, malononitrile and isatins were reacted in aqueous medium in presence of piperidine at room temperature.Rationalisation of reaction mechanism indicated the initial reaction of ethyl acetoacetate with hydrazine hydrate forming an intermediate pyrazine derivative, which further undergo Michael type of addition on another intermediate formed by the reaction of malononitrile and isatin. Further this reaction sequence is completed by the cyclisation and tautomerisation leading to the corresponding spiroheterocyclic derivatives. Further application of this protocol was explored by extending this reaction to acenaphthylene-1, 2-dione in place of isatin to afford spiro[acenaphthylene-1,4'-pyrano[2,3-c]pyrazole]-5'-carbonitrile in excellent yields. This elegant one pot synthesis, starting from readily accessible simple chemicals, provides an easy approach to attain biologically important spirooxindole derivatives by a green methodology.

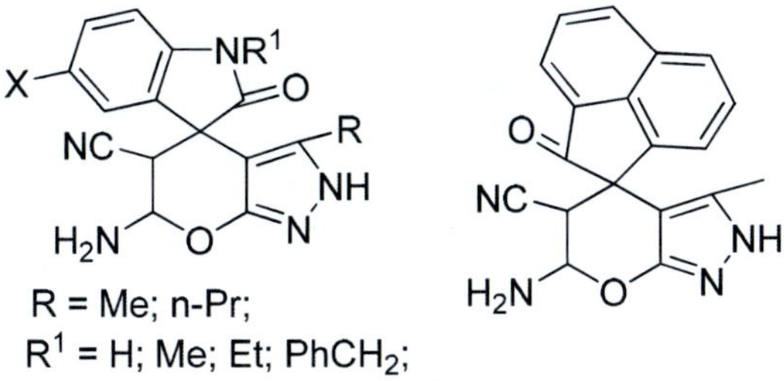

R = Me; n-Pr;
R^1 = H; Me; Et; $PhCH_2$;
X = H; Br; NO_2;

(III)

B. Imidazole Derivatives

Benzimidazole scaffold represents varied spectrum of biological activity and occurs in many pharmaceuticals as well as natural products. [18] Several synthetic methodologies are known for the preparation of benzimidazole derivatives, among which condensation of o-phenylene diamines and carboxylic acids (including their derivatives such as orthoesters,imidates or nitriles) was commonly used, involving harsh reaction conditions. Functional benzimidazoles can also be obtained by palladium catalysed tandem carbonylation –cyclisation of o-phenylenediamine [19], Pd catalysed dehydration, coupling reaction of 2-bromo aniline, [20] reductive cyclisation of o-nitroaniline with aldehydes, [21] rhodium catalysed hydroformylation of N-alkenyl phenylene diamines [22] as well as oxidative condensation- aromatization reaction involving o-phenylenediamines and aldehydes. [23] Yuanjiang pan et al recently investigated [24] the reaction between o-phenylene diamine and various aryl/heteroaryl aldehyde derivatives performed under aqueous phase conditions. The scope of this reaction was examined using different catalysts as well as different solvent systems, at a broad range of temparatures and it was observed that water in presence of TMSCl as a promoter resulted in remarkably better results in both product yield as well as selectivity point of view. Under optimized conditions, exclusively 1, 2-disubstituted benzimidazoles(IV) with various functional groups were obtained in appreciable yields and selectivity. Interestingly the scope of this water, TMSCl catalytic system, which is tolerant to varied functional groups, was extended to the synthesis of 2-substituted quinoxalines, which were prepared by the reaction of o-phenylene diamines and α-bromoketones, to obtain products in appreciable yields. This convenient methodology, where in water as a medium displayed inducing effect for the selective formation of 1,2-disubstituted benzimidazoles, is a very good contribution for aqueous phase organic synthesis.

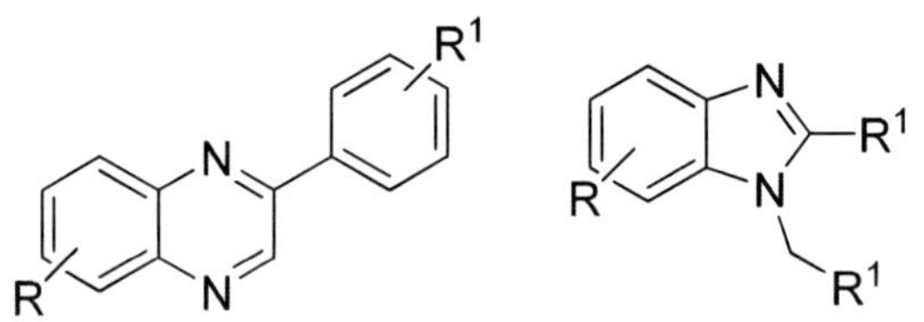

(IV)

C. Isoxazole Derivatives

Isoxazoles attracted attention due to their varied applications in the field of medicine, agriculture and industry. Isoxazole is a prominent structural skeleton widely occurring in medicinally important molecules, such as antidepressants, antipsychotics and antianxiolytics. Its derivatives are also used as building blocks in natural product syntheses. Certain isoxazole derivatives are known to posseses insecticidal, [25] antifungal, [26] antiepileptic [27] and anticonvulsant properties. [28] Generally, isoxazoles are synthesized by the insitu formation of nitrile oxide followed by intra molecular dipolar cycloaddition, [29] by the dehydration of nitro alkanes, [30] oxidation of aldoximes [31] or dehydrohalogenation of hydroximoyl chlorides, [32] for which a range of reagents like NaOCl, chloramines-T etc., can be used.

In a recent finding Mehdi Adib et al reported [33] an efficient novel synthesis of 5-(alkylamino)-4-aryl-3-isoxazol carboxamides(V) by the reaction of different nitro styrenes with isocyanides in water as a sole solvent at 80 °C in good yields. The simplicity of this approach for these target compounds, in which nitrogen and oxygen atoms of isoxazole ring are derived from the nitro group of the activated styrene molecule, unlike previously reported strategies as claimed by the authors, makes it an interesting contribution for isoxazole chemistry.

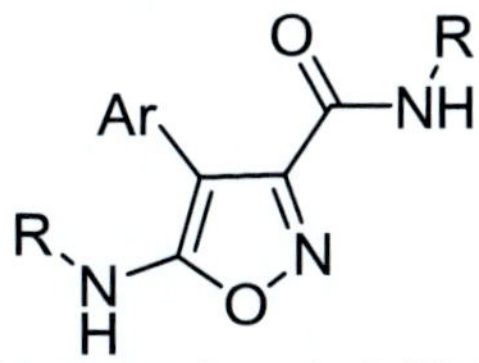

Ar = Substituted aryl; 2-Thienyl
R = Cyclohexyl; t-butyl; 1,1,3,3-tetramethylbutyl

(V)

In a recent publication Ching-Fa Yao et al described [34] the synthesis of certain tricyclic isoxazoline and isoxazole derivatives in water as a sole solvent medium, in which 2- allyoxy as well as 2- propargyloxy benzaldoximes underwent intramolecular cyclisation in presence of HTIB affording target compounds in good yields. The present methodology is a mild, eco friendly and economically viable alternate approach.

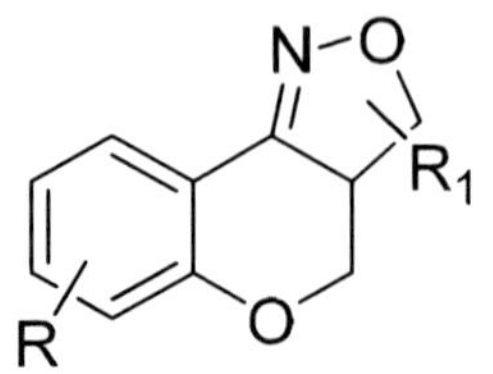

R = OCH_3; Cl; Br; NO_2; t-Butyl;
R_1 = H; CH_3; Ph; CH_2OAc;

D. Benzothiazole Derivatives

Benzothiazoles are known for their broad range of applications in the field of therapeutics, [35] enzyme inhibition, [36] plant growth regulation [37] and dyes [38]. In general, benzothiazoles can be prepared by the reaction of 2-amino thiophenol with an aromatic/hetero aromatic/ alkyl or aralkyl carboxylic acid, ester or amide, as well as by the intramolecular cyclization of ortho substituted aryl, alkyl thiourea derivatives. When 2-amino thiophenol is subjected to condensation with alkyl or aralkyl aldehyde, it affords benzo thiazoline moiety. In a traditional protocol, the methodology requires acidic catalysts like PPA, P_2O_5-$MeSO_3H$, or PTSOH adsorbed on SiO_2/K^{10} and graphite, or oxidizing agents like $K_3Fe(CN)_6$ or costly toxic catalysts as well as high boiling solvents. [39] In a recent research article, AK chakraborti et al described [40] an efficient, facile, convenient "on water" mediated synthesis of benzothiazoles/benzothiazolines(VI) in water as a sole solvent medium. In this interesting approach different substituted aromatic/ hetero aromatic, aldehydes are reacted with 2-amino benzene thiol in a one pot reaction at 110°C, affording 2-substituted benzothiazoles in excellent yields, where as alkyl/ aralkyl aldehydes gave benzothiazolines. The reaction, as reported by the authors is highly chemoselective with no other side reactions. Authors conducted representative experiment under different solvent conditions to bring out the potentiality of water as an efficient solvent in the present reactions. This highly useful green synthetic protocol is a significant contribution, as it avoids other catalysts/reagents, and toxic inflammable organic solvents.

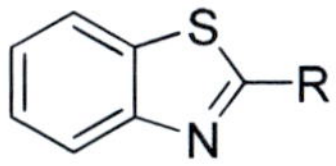

R = Aryl;Heteroaryl;
Aryl alkyl; Alkyl; NHMe;
(VI)

E. Thiadiazole Derivatives

1, 3, 4-Thiadiazoles are well recognized for their wide range of biological activities like herbicidal, [41]fungicidal, [42] bactericidal, [43] antiinflammatory [44] as well as carbonic anhydrase inhibitory property. [45] Also, they have useful applications in the field of dyes, [46] photographic materials, [47] optically active liquid crystals [48] and lubricating substances. [49] Among different methods available, cyclization of a 1,2-diacyl hydrazine or its thia analog in the presence of $SOCl_2$ or $POCl_3$ and strong mineral acid is a prominent protocol. [50] Other cyclising reagents in practice include Lawesson's reagent, TMSCL, and $(PhO)_2P(O)Cl$. Even though many multistep syntheses are available,single step syntheses for 1,3,4-thiadiazoles are few. First one pot, solvent free synthesis of substituted 1, 3, 4-thiadiazole derivatives by the reaction of acid hydrazides and triethyl orthoalkanoates was reported by Varma et al. [51] As an extension of their research in the area of green chemistry, M,R.Saidi et al described [52] an interesting preparation of 1,3,4-thiadiazoles derivatives(VII), reacting dithiocarbamate and acid hydrazides in presence of triethylamine or pyridine in aqueous medium. Scope and limitations of this methodology were explored using a broad range of dithiocarbamates prepared from aliphatic and aromatic primary amines and acid hydrazides. This simple one step, one pot ecofriendly procedure is an appreciable contribution to green chemistry.

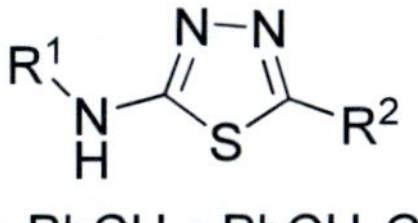

R^1 = Ph; $PhCH_2$; $PhCH_2CH_2$; n-bu; adamantyl;
R^2 = Ph; $PhCH_2$; 4-Pyridyl; p-Tolyl;
(VII)

F. Triazole Derivatives

1, 2, 4-Triazoles are widely acknowledged as a prominent class of heterocyclic system, due to their utility in agriculture, medicine, and metallo organics. [53] Many biologically active fused heterocyclic derivatives are prepared using 3(5)-amino-1, 2, 4-triazole as an important reaction intermediate. [54] In general 1, 2, 4-triazoles are synthesized by the cyclocondensation of amidoguanidines, in presence of bases or at high temperature involving complicated workup procedures. [55] Recently A.V.Dolzhenko et al described [56] an efficient, catalyst free synthesis of 3,5 amino-1,2,4-triazoles(VIII) by the cyclo condensation of (het)aryl amidoguanidines in water as sole solvent at reflux temperatures, resulting in quantitative yields of pure end products.

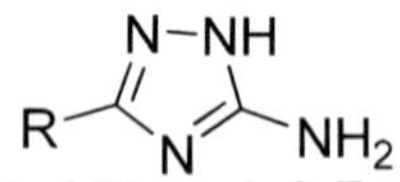

R = Substituted Phenyl; 2-Furyl; 2-Thienyl;
2-Pyridyl; 3-Pyridyl; 4-Pyridyl;

(VIII)

Six membered Heterocyclic Systems

A. Pyridine Derivatives

4-Substituted 1, 4-dihydropyridines have attracted attention due to their potent biological activity and are known to be drug candidates useful in the treatment of cardiovascular disorders. [57] These also act as calcium channel blockers. [58] Acridinediones consist of 1, 4-dihydropyridine as a parent nucleus and are important structural frameworks in heterocyclic chemistry.

In a recent report G.W.Wang etal described [59] the synthesis of unsymmetrical 4-arylacridinediones(IX) by an environ friendly one pot multicomponent approach. In this exhaustive study authors have excellently demonstated the efficacy of water as a sole solvent medium in bringing about cyclisation reaction. A mixture of equimolar amounts of enaminones, aromatic aldehydes, 1,3-cyclohexanone was reacted to yield the unsymmetrical acridinediones. The said reaction strategy was observed in varied reaction conditions such as neat conditions as well as in water medium at reflux temperatures, to demonstrate and prove the advantages of aqueous phase organic synthesis. It was observed that both in solvent free as well as water conditions products are obtained in higher yields than in organic solvent conditions. But under solvent free condition, reactions were to be conducted at higher temperatures where as lower temperature heating leads to the formation of undesired products. This methodology is an important contribution to green chemistry, due to its eco friendly nature, and simple work up procedures.

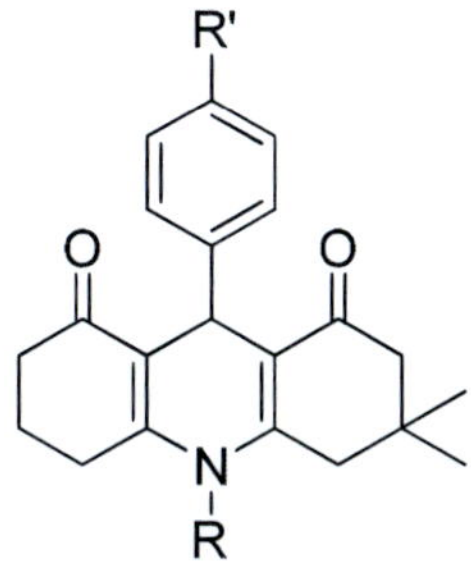

R = H; CH_3; 4-CH_3Ph;
R' = H; Cl; OCH_3
(IX)

Imidazo[1,2-a] pyridine moiety belongs to a prominent class of heterocyclic system widely recognized for their broad range of biological activity such as antibacterial, antifungal, antiviral, antiprotozoal, antihelmintic, antiulcer, anticonvulsant, antiinflammatory, hypnotic, anxiolytic and immunomodualtory properties. [60] The general methodogy for the synthesis of imidazo[1,2-a] pyridine system involves the reaction of an aldehyde, an isocyanide and 2-amino azine system. However in many of these classical protocols the role of a catalyst is inevitable. Montmorillonite K10, $Sc(OTf)_3$, perchloric acid or glacial acetic acid are some of the catalysts useful in the synthesis of imidazo[1,2-a] pyridines. However a recent report by Mehdi Adib etal described [61] a catalyst free three component, one pot reaction of 2-amino pyridines, aldehydes, and isocyanides in pure water as a sole solvent medium, affording 3-amino imidazo [1,2-a] pyridines(X) in quantitative yields. While standardizing and expanding the scope of the reaction various aromatic aldehydes were used in the reaction to broaden the substrate base. Apart from the synthesis of series of imidazo[1,2-a] pyridine derivatives, authors also applied this protocol for the preparation of imidazo[2,1-b][1,3] thiazoles, where in 2-amino thiazole was reacted with different aromatic aldehydes in presence of isocyanides in water at 70°C. This simple, convenient, catalyst free strategy is an important improvement to design and prepare a series of biologically active heterocyclic molecules.

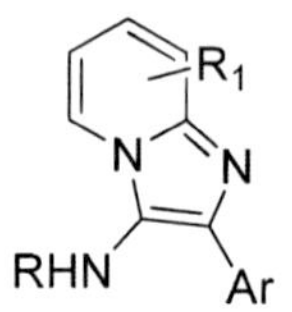

Ar = Phenyl; Substituted phenyl:
R = Cyclohexyl; Phenyl; t-Butyl;
R_1= H; CH_3;
(X)

B. Pyran Related Compounds

Pyran is an important structural motif widely occurring in a number of natural products [62] and photochromic materials. [63] It also exhibits potential biological activities of medical significance. [63] Classical synthetic approaches require various acidic reagents like phosphoric acid [64] as well as other catalysts such as $TiCl_4$, $In(OTf)_3$, BF_3-Et_2O, or EDDA/$ZnCl_2$. [65] In a recent research publication Yang Rok Lee described [66] an environmentally benign, efficient and simple synthesis of pyran derivatives(XI) by Domino Knoevenagel, electrocyclisation using water as a sole medium. By this 6π formal (3+3) cycloaddition approach authors demonstrated the potentiality of water as a solvent, extending the scope of this strategy to several biologically interesting structural scaffolds such as pyranocoumarins, pyranoquinolinones, and pyranonaphthaquinones along with certain other natural and non-natural products. The reported methodology is an important contribution in the field of green chemistry due to its operational simplicity and noninvolvement of metal catalysts.

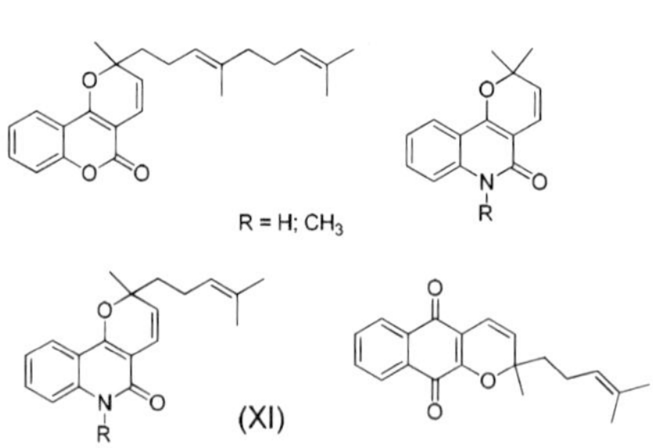

R = H; CH_3

(XI)

Coumarins and chromones have found wide range of applications [67] in the field of pharmaceuticals, agrochemicals, fragrances, cosmetics, food additives and optical brightening agents as well as dye lasers. These also exhibit diverse biological activities such as insecticidal, antihelmintic, hypnotic and anticoagulant properties. [68] In addition 3, 4-dihydro coumarins as well as thiochroman-4-ones and 1-thiochromone derivatives also represent prominent category of biologically active scaffolds. [69] In recent research publication, Moghaddam et al described [70] an efficient catalyst free synthesis of novel pentacyclic thiochromone-annulated thiopyrano coumarin derivatives(XII). The target compounds are prepared via Domino Knoevenagel-hetero-Diels-Alder reactions of 4-hydroxy dithiocoumarin and O-acrylated salicylaldehydes in water as sole solvent. One of the reactants i.e. O-acrylated salicylaldehydes were prepared from the corresponding salicylaldehydes by reacting with different acryloyl chlorides in presence of K_2CO_3 in dry acetone. Domino Knoevenagel-hetero-Diels-Alder strategy is a prominent synthetic methodology effectively practised to achieve hetero cyclic derivatives as well as natural products. By following this familiar approach, Moghaddam et al synthesized a series of pentacyclic thiopyrano coumarins in appreciable yields with high regio and stereo selectivity.

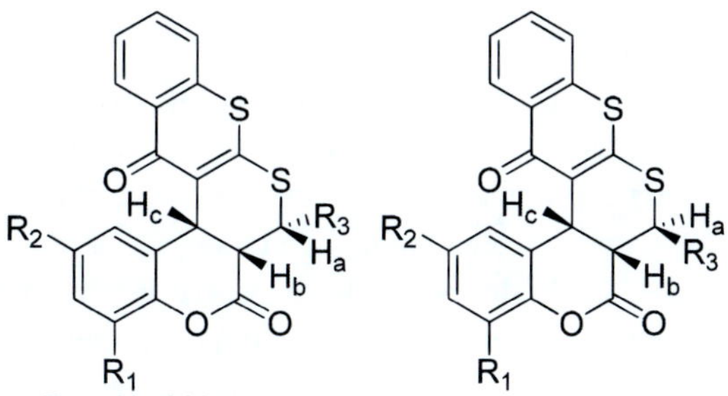

R_1 = H; OMe;
R_2 = H; Br; NO_2;
R_3 = Ph; Me;

(XII)

Dibenzoxanthene is an important core structural unit present in many biologically active compounds like antivirals, [71] antibacterial [72] and antiinflammatory agents, [73] apart from having application in the field of agriculture, dyes [74] and fluroscent materials. [75] General methods for the synthesis of benzoxanthenes include the reaction of β-napthol with formamide, 2-napthol -1- methanol and carbon monoxide. [76] In general these involve different types of catalysts and harsh reaction conditions. In recent research work Minoo Dabiri etal reported [77] a one pot ecofriendy protocol for the synthesis of alkyl or aryl- 14H- dibenzo[a,j]xanthenes(XIII) in aqueous medium in the presence of $KAl(SO_4)_2$ 12 H_20(Alum). While expanding the scope this study, they examined the efficacy of different solvents as well as catalysts on the synthesis of dibenzoxanthenes and observed that water as a sole solvent medium provided excellent yields of the cyclized products. Among different catalysts investigated, Alum i.e., $KAl(SO_4)_2$ 12 H_20, an inexpensive and readily available catalyst seemed to be the best choice. The generality of the reaction was established extending the reaction to different substituted aromatic and aliphatic aldehydes. This efficient, convenient, environ friendly protocol is valuable contribution for xanthene chemistry, with many advantages like use of water medium, easily available catalyst, and simple purification of the products by filtration technique, as well as shorter reaction times.

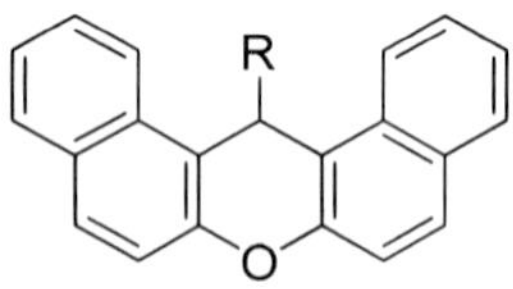

R = Phenyl; Substituted phenyl; CH_2CH_3; $(CH_3)_2CHCH_2$

(XIII)

Chromenes exhibit a wide range of biological activity such as spasmolytic, anticoagulant, diuretic, anticancer and antianaphylactic properties. [78] Isatin derivatives, having potential medicinal activities also act as precursors for several natural products. It was reported that existence of two or more different heterocyclic structures in a single molecule increases the biocidal activity. In this context it is worth mentioning, the studies of Minoo Dabiri et al, [79] in which isatin was reacted with malononitrile and 1, 3-dicarbonyl compounds, which resulted in a series of spiro chromenes(XIV) in water medium. The reaction was investigated under different conditions to evaluate the effect of various catalysts as well as solvent systems. Among several solvents examined reactions conducted in water medium resulted in excellent yields. Authors also examined the reaction in the absence and presence of various ammonium salts. It was observed that the reaction did not proceed in the absence of any catalyst. Among different ammonium salts studied as catalysts such as ammonium acetate, ammonium carbonate, ammonium sulfate, ammonium oxalate, ammonium sulfide and ammonium chloride, it was observed that ammonium chloride enchanced the yields to higher levels. To extend the scope of this protocol, reactions were conducted with variety of substances. Apart from malononitrile, ethylcyanoacetate also worked well. In place of isatin, acenapthoquinone also gave expected spiroproducts. Spiroacridines were obtained when isatin and 1, 3-cyclohexanedione were reacted with benzylcyanide in place of malononitrile. As explained by the authors, the reaction involves initial Knoevenagel reaction between isatin and CH-acidic substrate in presence of ammonium chloride in water followed by Michael addition of diketone to the unsaturated nitrile and the inter mediate undergoes cycloaddition forming the expected spiro product. This interesting one pot, three component water medium reaction is a potential addition to expand the scope of green chemistry.

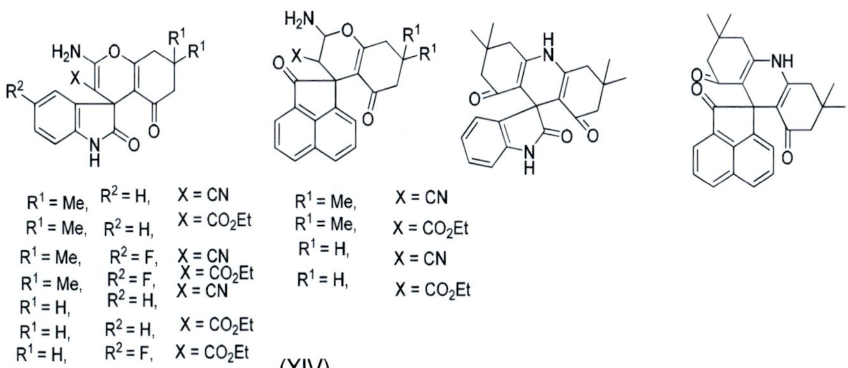

(XIV)

In view of various biological activities associated with pyrano[2,3-d], pyrido[2,3-d] pyrimidines, and pyrano[3,2-c]benzopyran derivatives, many attempts were made to synthesize these compounds involving multistep reactions and complex reaction pathways. However, Ahmad Shaabani etal developed [80] a one pot, three component condensation reaction resulting in these annulated heterocyclic systems (XV) in water medium. In this simple, uncatalysed interesting reaction, substituted aromatic aldehyde molecule undergoes Knoevenagel condensation reaction with alkyl nitrile producing 3-benzylidene-2,4-pentanedione, which in turn reacts with different active C-H acids resulting in a spectrum of products. Scope of this reaction has been extended to include 4-hydroxycoumarin, 4-hydroxy-6-methylpyrone, 1, 3-dimethylbarbituric acid and 1, 3-dimethyl-6-amino uracil as one of the three components. This green approach is a facile alternate protocol to the already existing methodologies to arrive at pyran annulated heterocyclic systems.

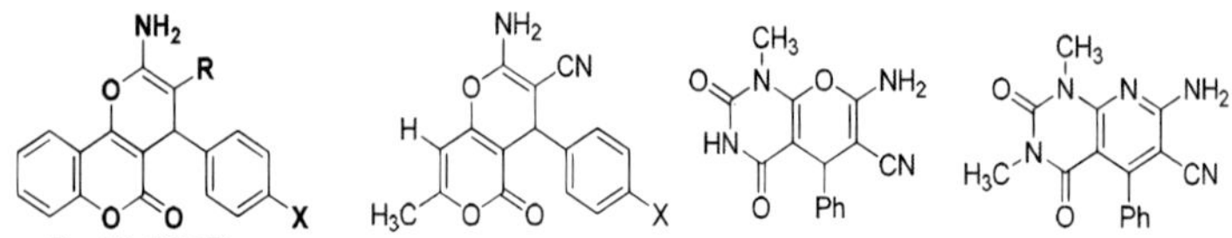

R = CN; COOEt
X = H; NO_2; OCH_3;

(XV)

C. Pyrimidine Derivatives

3, 4-Dihydropyrimidones belong to pharmacologically prominent class of compounds, having a wide spectrum of activity such as hypertensive, mitotic kinesin inhibition, α-adrenergic antagonism etc. [81] In addition N,N[1]-disubstitution enhances lipophilicity. [82]Eventhough due to their potential biological activity, several synthetic methods are available for 3,4-dihydropyrimidones, approaches towards N, N[1],dimethyl substituted 3,4-dihydro pyrimidones are relatively few.

I.Suzuki et al, utilizing the prominent Biginelli protocol, which is a widely used multicomponent reaction having larger applicability, reported the synthesis of diversely substituted 3, 4 dihydro-pyrimidin-2(1H)ones (XVI). [83] In this work, the titled scaffold was obtained by the reaction of β-ketoesters, aromatic aldehydes and urea in aqueous medium. Authors conducted an exhaustive study on the reaction, where in the catalytic efficiencies of metal triflates such as $Ni(OTf)_2$, $Cu(OTf)_2$ and $Yb(OTf)_3$, and metal triflimides like $Ni(NTf_2)_2$, $Cu(NTf_2)_2$, $Yb(NTf_2)_3$ and $Zn(NTf_2)_2$ were compared and described that in metal triflimides-catalyzed reaction prolonged reaction time improved the yields. Even though the metal triflates-catalyzed reaction afforded the desired compounds along with polymeric impurities, the metal triflimides did not form undesired side products and afforded expected 3, 4 dihydro-pyrimidin-2(1H)ones in excellent yields. This protocol conclusively proved that metal triflimide salts catalyzed Biginelli reaction in pure water more efficiently than metal triflates and the addition of Bronsted acids such as CH_3CO_2H, HCl etc, enhanced the yields in these reactions.

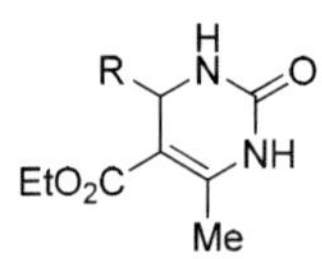

R = Phenyl; Substituted Phenyl; n-Pentyl;

(XVI)

Mukhopadhyay et al in a recent research work described [84] an efficient, three component one pot protocol for the synthesis of N-N'-dimethyl substituted 4-aryl-3, 4-dihydro pyrimidines(XVII) in water medium, catalysed by $MgSO_4$. While optimizing the reaction conditions, different alkaline metal(II) sulfates like $CaSO_4$ $BaSO_4$, $SrSO_4$, and $MgSO_4$ were examined for their efficacy as catalysts and among which $MgSO_4$ proved to be the best choice.

Initially, ethylacetoacetate was reacted with 4-chlorobenzaldehyde and urea. While establishing the reaction parameters, different solvents like H_2O, CH_2Cl_2, THF, MeOH, EtOH, DMF, DMSO were studied for their utility and H_2O was proved to be affording better results. Moreover, authors extended the scope of this reaction to a wide variety of substrates both in the aldehyde part as well as 1, 3-dicarbonyl compounds. Both β-diketones and β-ketoesters were used as substrates. Urea and thiourea were equally effective in affording the corresponding products. Acylimine intermediate, which formed initially, reacts with β-diketones or β-ketoesters subsequently to complete the sequence of reactions affording expected products. Authors explained that Mg^{+2} helps in forming strong metal oxygen bond, which helps in increasing electrophilicity of the β-diketone or β-ketoester leading to cyclisation and dehydration to obtain dihydropyrimidones. The reported methodology, which is an excellent addition to green chemistry, has the advantages of conducting reaction in water medium, with a simple, affordable, easily available catalyst $MgSO_4$, apart from easy isolation of the products by simple filtration techniques.

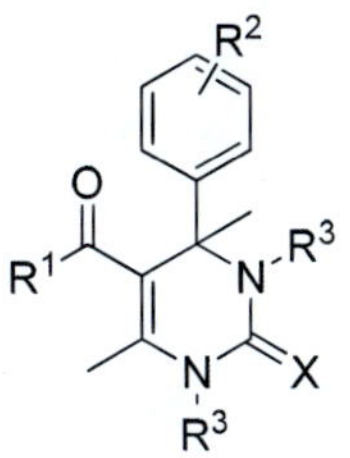

R^1 = Me; OMe; OEt;
R^2 = Phenyl; Substituted Phenyl; 2-Furanyl;
R^3 = H; Me;
X = O; S;

(XVII)

D. Quinazoline Derivatives

Octahydroquinazolinones exhibit antibacterial activity against Staphylococcus aureus, Escherichia coli and Pseudomonas aeruginosa. [85] Generally these can be prepared by the reaction of aldehydes with $SOCl_2$ and pyridine, and then with 2-aminobenzylamine, which involve harsh reaction conditions and expensive hazardous catalysts. In a research report, M. R. Islami et al described [86] an interesting synthesis of Octahydroquinazolinone and Octahydrothioquinazolinone(XVIII) derivatives, by a simple multicomponent one pot Biginelli reaction in which urea/thiourea, dimedone, and aromatic aldehydes are reacted in aqueous medium, containing few drops of sulfuric acid resulting in quantitative yields.

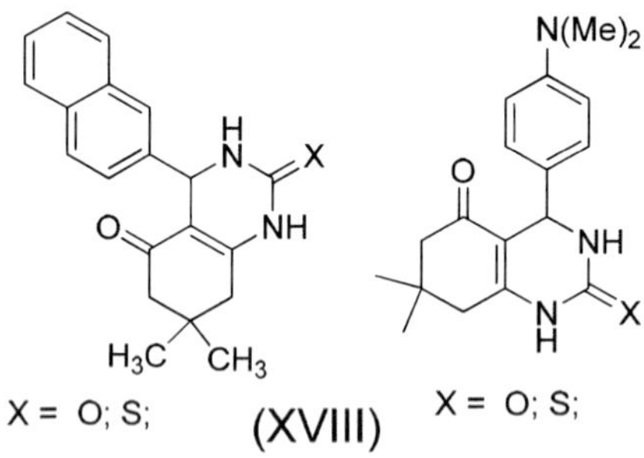

(XVIII)

E. Benzoxazines and Benzothiazines

Nitrogen heterocyclic systems are extensively used as prominent and potential building blocks in the synthesis of various pharmaceutical compounds. Benzoxazines display a broad range of biological activities such as antihypertensive, [87] antirheumatic and potassium channel modulation. [88] These act as intermediates in organic syntheses, particularly in peptide synthesis. [89] Several efforts have been made to develop benzoxazines from 2-aminophenol by different annulation methods, [90] involving either Palladium-catalysis [91] or use of microwaves [92] as well as bases. [93] A facile, efficient and green synthetic protocol for the synthesis of benzofused N,S-, N,O- and N,N-hetero cycles(XIX) was described by Xian-Fu Lin et al. [94] By this simple methodology, which involved readily available reactants and milder reaction conditions, diverse 2-aminophenols, 2-aminothiophenols and 1, 2 phenylenediamines were reacted with DMAD/DEAD to obtain various benzoxazines, benzothiazine and quinoxaline derivatives respectively, using water as sole solvent at 50°C, in almost quantitative yields. Authors in their investigative studies observed the effect of various solvents on this cyclization reaction and proved the efficacy of water over others. Product formation was also found to be influenced by the nucleophilicty of the heteroatom, which was in the order S>O>N. The present protocol is an interesting contribution for the development of green chemistry.

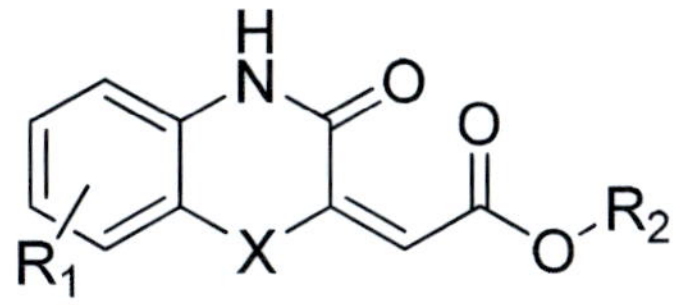

X = S, O, NH
R_1 = Cl; NO_2; t-Bu; CH_3;
R_2 = CH_3; C_2H_5;

(XIX)

F. Quinoxaline Derivatives

Quinoxalines and their derivatives are prominent among nitrogen containing heterocyclic systems as they exhibit wide range of biological activities such as antifungal, antibacterial, antiviral, insecticidal, antihelmintic as well as anticancer properties. [95] Despite the presence of a number of synthetic approaches available, research and design of a convenient, environmentally benign protocol is always desirable to expand the scope of green chemistry. They are important building blocks for the synthetic organic semiconductors, dyes and electroluminiscent materials. [96] They also act as useful sub units in macrocyclic receptors for molecular recognition and chemically controllable switches. [97] Moreover these molecules also serve as platforms for diversity oriented synthesis in solid phase. [98] The general methods for the preparation of quinoxaline derivatives include condensation of 1,2-diamines and 1,2-dicarbonyl compounds, [99] oxidative cyclization of α-hydroxy ketones with 1,2-diamines, [100] 1,4 addition of 1,2-diamines to diazenylbutenes, [101] cyclization-oxidation of phenacyl bromides with 1,2-diamines by $HClO_4.SiO_2$. [102]and oxidative coupling of epoxides with ene-1,2-diamines. [103] Pyrrolo-[1.2 a] quinoline scaffold occurs widely in natural products as well as pharmaceuticals. In a well conducted exhaustive investigation C.M.Che et al, examined [104] the reaction of several alkynyl substituted arylamines and aromatic alkynes, under different solvent systems as well as catalytic species. Among the solvents tested water afforded encouraging results. While expanding the substrate scope, a broad range of 1,4- amino alkynes, substituted with several steric as well as electronic properties were reacted under this protocol, affording diversely substituted pyrrolo[1,2-a]quinolines(XX). Among the catalytic systems, Au^{I}catalyst ie. $[Au\{P(tBu)_2$ (o-biphenyl)$\}]Cl/Ag\ SbF_6$ furnished very good yields. This simple and efficient methodology for the synthesis of highly substituted pyrrolo[1,2-a]quinolines, wherein regio and chemo selectivity during alkyne C-H activation was well controlled, affords several novel tri and tetracyclic amines, in which various C-C and C-N bonds were formed in a single step reaction.

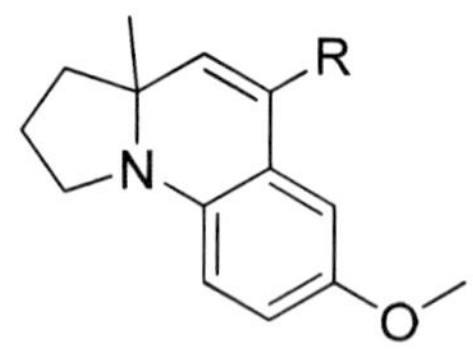

R = Phenyl; Substituted Phenyl; n-butyl;

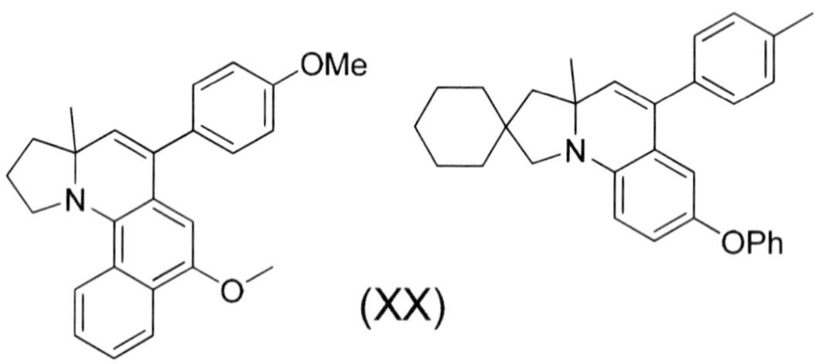

(XX)

Nageswar et al in a recent report described [105] a facile catalyst free synthesis of a wide variety of 3-substituted quinoxalinone(XXI) derivatives, which had the advantage of rapid product isolation and operational simplicity precluding hazardous organic solvents. In this protocol several benzene-1, 2-diamines were subjected to cyclocondensation reaction with several pyruvate derivatives in aqueous medium at 50°C. The scope of this methodology was also extended to certain α-keto esters also, affording the corresponding products in excellent yields. The plausible mechanism was explained as involving inter molecular condensation of benzene-1, 2-diamine with α-keto group of the ester, producing the intermediate ethyl 2-[(2-aminophenyl)imino]propanoate. This intermediate later undergoes intra molecular cyclization leading to the quinoxalinone derivatives.

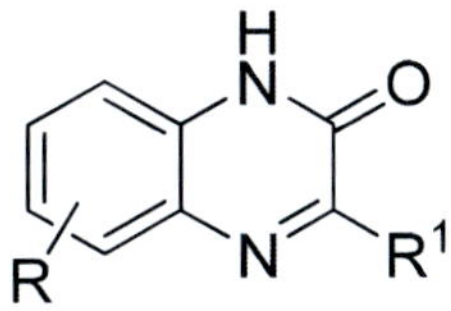

R = dichloro; dimethyl;

$R^1 = CH_3$; Ph; CF_3; $(CH_3)_2C\text{-}CH_2OH$

(XXI)

Yong Rok Lee et al in a recent publication described [106] the synthesis of 3,4-dihydro quinoxaline-2-amines(XXII) by a one pot three component method involving the reaction of o-phenylene diamines, different carbonyl compounds and isocyanides in aqueous medium catalysed by EDTA. The investigation covered the effect of different Bronsted acid catalysts on this reaction in water medium. Authors suggest that here EDTA acts as a Bronsted acid catalyst. Iminium cation formation is facilitated by EDTA with the protonation of carbonyl group, followed by nucleophilic addition of isocyanide to the iminium intermediate, followed by intramolecular cyclisation and isomerisation affording the 3, 4-dihydroquinoxaline-2-amines in excellent yields. This ecofriendly synthetic protocol is also claimed to be useful for scaleup preparations as well as structurally unique spirocyclic compounds.

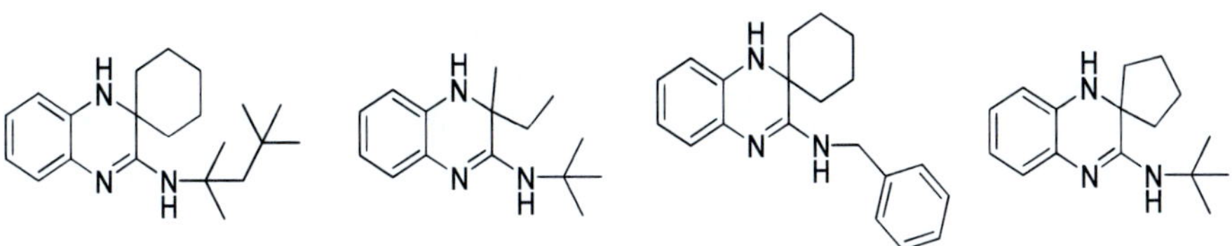

(XXII)

G. Indolizine Derivatives

Indolizine, a prominent N-bridgehead bicyclic ring system fused with both π-deficient pyridine and a π-excessive pyrole has attracted the attention, [107] due to their important biological activity with a variety of applications in medicinal chemistry. [108] Metal catalysed C-N bond forming approach for the synthesis of indolizines has been widely acknowledged for its efficacy as well as the wider scope for a broad range of applications. [109] Cycloisomerisation of alkynyl pyridines, [109] Cu-catalysed cycloisomerisation of OAc-substituted propargylic pyridine derivatives, [110] Pd-Cu catalysed one-pot synthesis of 3-amino indolines through cascade coupling / cycloisomerisation of propargyl amines or amides with heteroaryl bromides [110] and Au-catalysed 1,2-migration / cycloisomerisation of propargylic compounds [109] are some of the methodologies reported for the construction of indolizine molecule framework. In a recent study Liu et al investestigated [111] a gold (lll) Catalysed multicomponent coupling-cycloisomerisation reaction of heteroaryl aldehydes, amines and alkynes, which was conducted both in aqueous phase as well as solvent free conditions. During optimisation, authors examined the effect of various copper and gold catalysts on this multi component one pot interesting reaction, in which pyridine-2-carboxaldehyde was reacted with several amines as well as arylalkynes resulting in the formation of 1-3 disubstituted indolizine derivatives(XXIII) in excellent yields. Moreover, authors also extended the scope of this approach to include amino acid derivatives as useful building blocks in place of amino compounds providing N-indolizine incorporated α-aminoacids with high enantiomeric purity. The reaction is supposed to proceed via the formation of NR^1R^2-substituted propargylic pyridine scaffold via a Mannich-Grignard reaction. This methodology offers a quick access to substituted amino indolizines with high catalytic efficiency as well as atom economy.

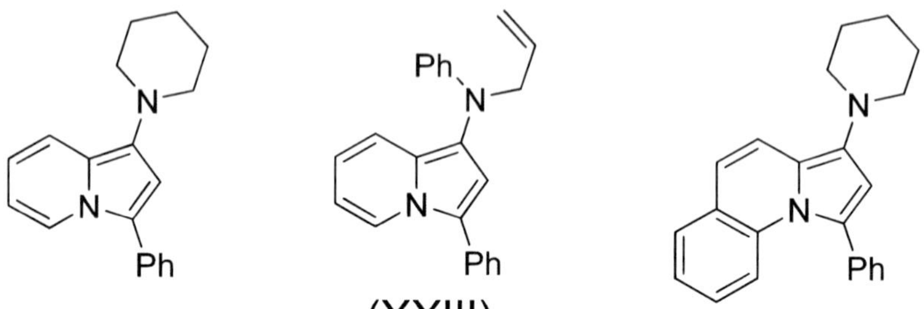

(XXIII)

Very recent research contribution for green chemistry by Richmond Sarpang et al describes [112] for the first time the efficiency of water as a sole solvent medium for the hetero cyclo isomerisation of pyridyl propargyl alcohols at reflux temperatures, affording the end products(XXIV) in quantitative yields. In this article authors brought out a comparative study, focusing on the potentiality of water as an excellent medium, whereby different metal catalysts, organic solvents, as well as other additives can be avoided, as a remarkable improvement in Lewis acid-catalyzed isomerization reactions, bringing out a major advantage of following green chemistry guidelines.

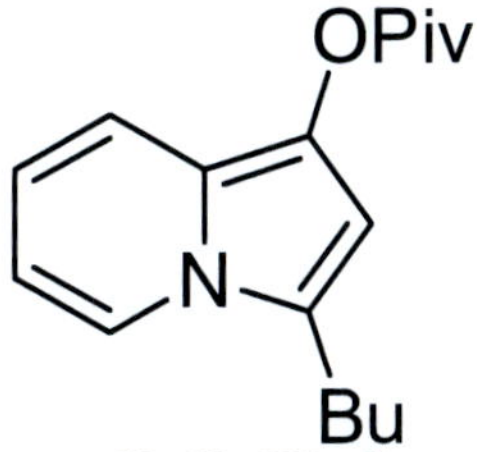

(XXIV)

SEVEN MEMBERED HETEROCYCLIC SYSTEMS

A. Azepine Derivatives

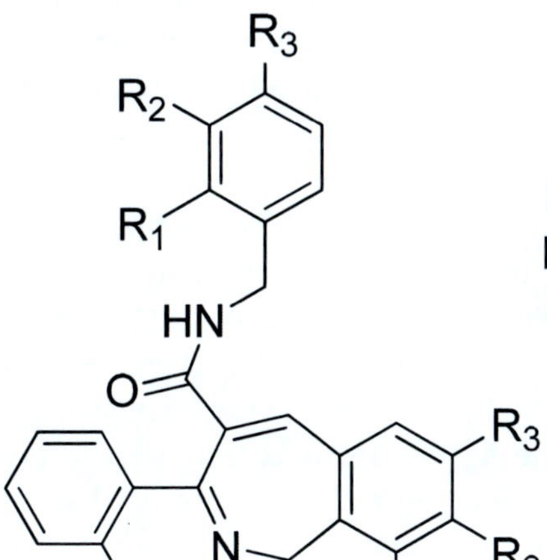

R_1 = H; CH_3; Cl;
R_2 = H; CH_3; OCH_3;
R_3 = CH_3; OCH_3; Cl; F;

(XXV)

Azepine is the important structural frame work widely present in many biologically active compounds as well as natural products. [113] Among these benzazepines attracted attention globally due to their potential pharmacological properties. [114] Accordingly, significant volume of research work was carried out leading to the development of varied synthetic strategies for the preparation of benzazepines. [115] Apart from the classical model approaches, recently described novel and eco friendly catalyst free domino process for the synthesis of several 2-benzazepines derivatives(XXV) in water as solvent, by Anil Kumar etal, is worth mentioning. [116] In this straightforward one pot preparation 4-chloro-3-formyl-coumarin was reacted with several substituted benzyl amines without any catalyst in water at room temperature affording substituted 5-(2'-hydroxyaryl)-4-amido-2- benzazepines in high yields, where in pure products are directly isolated by simple filtration.

B. Diazepine Derivatives

Benzodiazepine is an important structural motif responsible for a wider spectrum of biological activities including anti-inflammatory, analgesic, antidepressant, hypnotic, sedative and anticonvulsant properties. [117] Among these benzodiazepine derivatives, 2,4-disubstitution allow rapid construction of poly heterocyclic systems based on diazepine resulting in diversified libraries of small molecules of pharmacological significance. [118] In view of the above, the recent research finding by K.V. Srinivas et al is worth mentioning. [119] Authors described an efficient novel one pot, three component synthesis of 2,4-di substituted 3H-benzo[b][1,4] diazepines(XXVI) in aqueous medium, involving the formation of ynones followed by the in situ Michael addition and cyclo condensation of o-phenylene diamines. In this synthetic protocol benzoyl chloride was reacted with phenyl acetylene using $Pd(OAc)_2$ as a catalyst in presence of triethyl amine as a base, under copper, ligand and solvent free conditions, resulting in the formation of ynone at room temperature. This was further reacted insitu with o-phenylene diamines in water medium affording the benzodiazepine derivative. Authors extended the scope of this protocol by subjecting several substituted benzoyl chlorides, as well as phenyl acetylenes to this reaction strategy. Following the same approach target molecules were obtained in excellent yields, when heteroaryl acid chlorides were reacted. Variety of benzodiazepine molecule libraries can be prepared using this simple novel and convenient methodology, which extends the scope of water medium reaction for the growth of heterocyclic chemistry.]

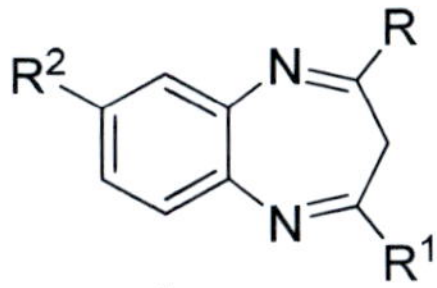

$R = R^1 = H; CH_3;$

R^2 = Phenyl; Substituted Phenyl;

(XXVI)

CONCLUSION

The expanding arena of aqueous medium organic synthesis showcases the significance of water as a medium to conduct organic reactions for the benefit of environmental sustainability by reducing the application of toxic organic solvents towards the reaction processes and further research in this field will contribute to attain newer goals.

ACKNOWLEDGMENTS

We thank CSIR, New Delhi for awarding fellowships to K.H.V.R and S.N.M.

REFERENCES

[1] Rideout, D. C.; Breslow, R. *J.Am. Chem.Soc.* 1980. 102. 7816-7817.

[2] Grieco, P. A.; Organic Synthesis in Water, Blacky Academic and Professional, London, 1998; Li, C.J.; Chan, T.H. *Organic Reactions in Aqueous Media,* Wiley & Sons, New York, 1997.

[3] Lindstrom, U. M. *Chem.Rev*. 2002, 102, 2751.

[4] Itami, K.; Yoshida, *J.-I. Chem.Rec.* 2002. 2. 213

[5] Lipshutz, B. H.; Aguinaldo, G. T.; Ghorai, S.; Voigtritter, K. Org. Lett. 2008, 10, 1325.; Lipshutz, B. H.; Chung, D. W.; Rich, B. Org. Lett. 2008, 10, 3793.; Lipshutz, B. H.; Ghorai, S.; Aguinaldo, G. T. AdV. Synth. Catal. 2008, 350, 953.; Lipshutz, B. H.; Taft, B. R. Org. Lett. 2008, 10, 1329.; Lipshutz, B. H.; Abela, A. R. *Org. Lett.* 2008, 10, 5329.

[6] Molteni, V.; Hamilton, M. M.; Mao, L.; Crane, C. M.; Termin, A. P.; Wilson, D. M. Synthesis 2002, 1669; Ferro, S.; Rao, A.; Zappala, M.; Chimirri, A.; Barreca, M. L.; Witvrouw, M.; Debyser, Z.; Monforte, P. Heterocycles 2004, 63; Ohberg, L.; Westman, J. Synlett 2001, 1296; Kidwai, M.; Saxena, S.; Khan, M. K. R.; Thukral, S. S. Bioorg. Med. Chem. Lett. 2005, 15, 4295; Hua, G. P.; Tu, S. J.; Zhu, X. T.; Zhan, X. J.; Xu, J. N.; Zhang, J. P.; Shi, F.; Wang, Q.; Ji, S. *J. Chin. J. Chem.* 2005, 23, 1646.

[7] Li, C. J.Chem Rev 1993. 93. 2023. Li, C. J.Chem Rev 2005. 105. 1273. Li, C.J. Chem Rev. 2005. 105. 3095. Li, C. *J. Chen, L. Chem. Soc.Rev.* 2006, 35. 68.

[8] Tripathy, R.; Ghose, A.; Singh, J.; Bacon, E. R.; Angeles, T. S.; Yang, S. X.; Albom, M. S.; Aimone, L. D.; Herman J. L and Mallamo, J. P. Bioorg. Med. Chem. Lett., 2007, 17, 1793; Sheldon, C. X.; Anthony, G. R.; Bheema, P. R.; Jeffery, S. A.;.Hasanthi, W. P. US PatentWO 2006/023931 A2; Singh, J.; Tripathy, R. *US Patent* WO 2001/32653 A1.

[9] Kemnitzer, W.; Drewe, J.; Jiang, S.; Zhang, H.; Zhao, J.; Crogan- Grundy, C.; Xu, L.; Lamothe, S.; Gourdeau, H.; Denis, R.; Tseng, B.; Kasibhatla, S and Xiong Cai, S. J. Med. Chem., 2007, 50, 2858; Kasibhatla, S.; Gourdeau, H.; Meerovitch, K.; Drewe, J.; Reddy, S.; Qiu, L.; Zhang, H.; Bergeron, F.; Bouffard, D.; Yang, Q.; Herich, J.; Lamothe, S.; Xiong Cai, S and Tseng, B. Mol. Cancer Ther., 2004, 3, 1365; Gourdeau, H.; Leblond, L.; Hamelin, B.; Desputeau, C.; Dong, K.; Kianicka, I.; Custeau, D.; Bourdeau, C.; Geerts, L.; Xiong Cai, S.; Drewe, J.; Labrecque, D.; Kasibhatla, S and Tseng, B. *Mol. Cancer Ther.,* 2004, 3, 1375.

[10] Bonsignore, L.; Loy, G.; Secci D and Calignano, A. *Eur. J. Med. Chem.,* 1993, 28, 517.

[11] Kumaravel, K.; Vasuki, G. *Green Chem.* 2009, 11, 1945-1947.

[12] Junek, H.; Aigner, H. Chem. Ber. 1973, 106, 914.; Wamhoff, H.; Kroth, E.; Strauch, K. Synthesis 1993, 11, 1129; Tacconi, G.; Gatti, G.; Desimoni, G. J. Prakt. Chem. 1980, 322, 831; Sharanina, L. G.; Marshtupa, V. P.; Sharanin Yu, A. Khim. Geterosikl. Soedin. 1980, 10, 1420; Sharanin Yu, A.; Sharanina, L. G.; Puzanova, V. V. *Zh. Org. Khim.* 1983, 19, 2609.

[13] Vasuki, G.; Kumaravel, K. *Tetrahedron Letters.*2008, 49, 5636–5638.

[14] Joshi, K. C.; Chand, P. Pharmazie 1982, 37, 1.; Da-silva, J. F. M.; Garden, S.J.; Pinto, A. C. J Braz Chem Soc 2001, 12, 273.; Abedl-Rahman, A. H.; Keshk, E. M.; Hanna, M. A.; El-Bady, Sh. M. *Bioorg Med Chem* 2006, 12, 2483.

[15] Kang, T.-H.; Matsumoto, K.; Murakami, Y.; Takayama, H.; Kitajima, M.; Aimi, N.; Watanabe, H. *Eur J Pharmacol* 2002,444, 39.; Ma, J.; Heht, S. M. Chem Commun 2004, 1190.; Khafagy, M. M.; El-Wahas, A. H. F. A.; Eid, F. A.; El-Agrody, A. M. Framaco 2002, 57, 715.

[16] Joshi, K. C.; Jain, R.; Sharma, K. J. Indian Chem Soc 1988, 65, 202.; Joshi, K. C.; Jain, R.; Arora, S. J. Indian Chem Soc 1988, 65,277.; Higashiyama, K.; Otomasu, H. *Chem Pharm Bull* 1988, 28, 648.

[17] Ahadi, S.; Yasaei, Z.; Bazgir, A. *J. Heterocyclic.Chem.* 2010, 47, 1090-1094.

[18] Preston, P. N.; Chem. Rev., 1974, 74, 279; Al Muhaimeed, H.; J. Int. Med. Res., 1997, 25, 175; Scott, L. J.; Dunn, C. J.; Mallarkey, G.; Sharp, M. Drugs, 2002, 62, 1503; Venkatesan, P. *J. Antimicrob. Chemother.,* 1998, 41, 145.

[19] Perry, R. J.; Wilson, B. D. *J. Org. Chem.,* 1993, 58, 7016.

[20] Brain, C. T.; Brunton, S. A. *Tetrahedron Lett.,* 2002, 43, 1893.

[21] Yang, D. L.; Fokas, D.; Li, J. Z.; Yu, L. B.; Baldino, C. M. *Synthesis,* 2005, 47.

[22] Anastasiou, D.; Campi, E. M.; Chaouk, H.; Jackson, W. R. *Tetrahedron,* 1992, 48, 7467.

[23] Trivedi, R.; De, S. K.; Gibbs, R. A. J. Mol. Catal. A: Chem., 2006, 245, 8; Beaulieu, P. L.; Hache, B.; Von Moos, E. Synthesis, 2003, 1683; Bahrami, K.; Khodaei, M. M.; Kavianinia, I. Synthesis, 2007, 547; Baharami, K.; Khodaei, M. M.; Naali, F. J. Org. Chem., 2008, 73, 6835; Sharghi, H.; Aberi, M.; Doroodmand, M. M. Adv. Synth. Catal., 2008, 350, 2380; Chen, Y. X.; Qian, L. F.; Zhang, W.; Han, B. Angew. Chem., Int. Ed., 2008, 47, 9330; Saha, D.; Saha, A.; Ranu, B. C. *Green Chem.,* 2009, 11, 733.

[24] Wan, J. P.; Gan, S. F.; Wu, J. M.; Pan, Y. *Green Chem.* 2009, 11, 1633-1637.

[25] Wickiser, D. I. U.S. Patent 4,336,264, 1982; *Chem. Abstr.* 1982, 96, 162684k.

[26] Romagnoli, C.; Vicentini, C. B.; Mares, D. *Lett. Appl. Microbiol.* 1995, 20, 5–6; Raffa, D.; Daidone, G.; Maggio, B.; Schillaci, D.; Plescia, F.; Torta, L. Il Farmaco 1999, 54, 90–94.

[27] Goel, A.; Madan, A. K. *Struct. Chem.* 1997, 8, 155–159.

[28] Martin, S. W.; Bishop, F. E.; Kerr, B. M.; Moor, M.; Moore, M.; Sheffels, P.; Rashed, M.; Slatter, J. G.; Berthon-Cédille, L.; Lepage, F.; Descombe, J. J.; Picard, M.; Baillie, T. A.; Levy, R. H. *Drug Metab. Dispos.* 1997, 25, 40–46.

[29] Jaeger, V.; Colinas, P. A. 'Nitrile Oxides', in Synthetic Applications of 1,3-Dipolar Cycloaddition Chemistry Toward Heterocycles and Natural Products, ed. Padwaa, A.; Pearson, W. H.; *Chemistry of Heterocyclic Compounds,* Wiley, Hoboken, 2002, vol. 59, p. 361.

[30] Gao, S.; Tu, Z.; Kuo, C.W.; Liu, J.T.; Chu C.M and Yao, C.F. Org. Biomol. Chem., 2006, 4, 2851; Tu, Z.; Jang, Y.; Lin C.; Liu, J.T.; Hsu, J.; Sastry, M. N. V.; Yao, C.F. Tetrahedron, 2005, 61, 10541; Liu, J.T.; Lin, W.W.; Jang, J.J.; Liu, J.Y.; Yan, M.C.; Hung, C.; Kao, K.H.; Wang, Y. Yao, C.F. Tetrahedron, 1999, 55, 7115; Yan M.C.; Liu, J.Y.; Lin, W.W.; Kao, K.H.; Liu, J.T.; Jang, J.J. Yao, C.F. *Tetrahedron,* 1999, 55, 12493.

[31] Becker N.; and Carreira, E. M.; Org. Lett., 2007, 9, 3857; Perez, L.; Khouly, M. E. E.; Cruz, P. D. L.; Araki, Y.; Ito. O.; Langa, F. Eur. J. Org. Chem., 2007, 2175; Nivlet, A.; Dechoux, L.; Martel, J. P.; Proess, G.; Mannes, D.; Alcaraz, L.; Harnett, J. J.; Gall, T. L. Mioskowski, C. Eur. J. Org. Chem., 1999, 3241; Sammelson, R. E.; Miller, R. B. Kurth, M. J. *J. Org. Chem.,* 2000, 65, 2225.

[32] Fraefel, N. L.; and Carreira, E. M.; Chem. Eur. J., 2009, 15, 12065; Domingo, L.R.; Chamorro, E.; and erez P. P.; Eur. J. Org. Chem., 2009, 3036; Singh VSingh., V. and Batra, S.; Eur. J. Org. Chem., 2008, 5446; Grecian, S. and Fokin, V. V.; Angew. Chem., Int. Ed., 2008, 47, 8285; Herrera, R.; Mendoza J. A.; Morales M. A. endez, F. M. Jim´enez-V´azquez, H. A. Delgado F and. Tamariz, J Eur. J. Org. Chem., 2007, 2352; Yamamoto H., Hayashi, S. Kubo M.,. Harada M,. Hasegawa M,. Noguchi M,. Sumimoto M and Hori, K.Eur. J. Org. Chem., 2007, 2859; Barr, L. Lincoln S. F and. Easton, C. J Chem. Eur. J., 2006, 12, 8571; Giorgi, G.. Lampariello L. R, Minetto G., Paoli M. L. Riello, V. Rodriquez M and. Sega A, Eur. J. Org. Chem., 2003, 4777; Roussel, C Fihi, R. Ciamala K. Audebert, P and. Vebrel, J New J. Chem., 2000, 24, 471; Adembri, G.; Giorgi G,. Lampariello R. L, Paoli M. L. and Sega, A.; *J. Chem. Soc., Perkin Trans.* 1, 2000, 2649.

[33] Adib, M.; Mahdavi, M.; Ansari, S.; Malihi, F.; Zhu, L. G.; Bijanzadeh, H.R. *Tetrahedron Lett.* 2009, 50, 7246-7248.

[34] Raihan, M. J.; Kavala, V.; Kuo, C.W.; Raju, B. R. Yao, C.F. *Green Chem.* 2010, 12, 1090-1096.

[35] Yoshida, M.; Hayakawa, I.; Hayashi, N.; Agatsuma, T.; Oda, Y.; Tanzawa, F.; Iwasaki, Koyama, S. K.; Furukawa, H.; Kurakata, S. and Sugano, Y. *Bioorg. Med. Chem. Lett.,* 2005, 15, 3328.

[36] Choi, S.-J.; Park, H. J.; Lee, S. K.; Kim, S. W.; Han, G. and Choo, H.-Y. P. *Bioorg. Med. Chem.,* 2006, 14, 1229.

[37] Loos, D.; Sidoova, E.and Sutoris, V.; *Molecules,* 1999, 4, 81.

[38] Razus, A. C.; Birzan, L.; Surugiu, N. M.; Corbu, A. C. and Chiraleu, F. *Dyes Pigm.,* 2007, 74, 26.

[39] Itoh, T. and Mase, T. Org. Lett., 2007, 9, 3687; Evindar, G. and Batey, R. A. J. Org. Chem., 2006, 71, 1802; Heo, Y.; Song, Y. S.; Kim, B. T. and Heo, J.-N. Tetrahedron Lett., 2006, 47, 3091; X.-J. Mu, J.-P. Zou, R.- Zeng, S. and Wu, J.-C. Tetrahedron Lett., 2005, 46, 4345; Rostamizadeh, S. and Housaini, S. A. G.; Phosphorus, Sulfur Silicon Relat. Elem., 2005, 180, 1321; Rudrawar, S. ;Kondaskar, A. and Chakraborti, A. K. Synthesis, 2005, 2521, Chakraborti, A. K.; Selvam, C.; Kaur, G. and Bhagat, S. Synlett, 2004, 851; Chakraborti, A. K.; Rudrawar, S.; Sharma, L. and Kaur, G.; Synlett, 2004, 1533; Zificsak, C. A.; and Hlasta, D. J.; *Tetrahedron,* 2004, 60, 8991.

[40] Chakraborti, A. K.; Rudrawar, S.; Jadhav, K.B.; Kaur, G. and Chankeshwara, S. V. *Green Chem.* 2010, 9, 1335-1340.

[41] Jin, G. Y.; Hou, Z.; Zhao, G. F.; Cao, C. Y.; Li, Y. C. *Chem. J. Chin. Univ.* 1997, 18, 409–412.

[42] Lu, S. M.; Chen, R. Y. *Org. Prep. Proced. Int.* 2000, 32, 302–306.

[43] Hui, X. P.; Zhang, L. M.; Zhang, Z. Y.; Wang, Q.; Wang, F. *Indian J. Chem., Sect. B* 1999, 38, 1066–1069

[44] Kumar, H.; Javed, S. A.; Khan, S. A.; Amir, M. Eur. J. Med. Chem. 2008, 43, 2688–2698; Mullican, M. D.; Wilson, M. W.; Connor, D. T.; Kostlan, C. R.; Schrier, D. J.; Dyer, R. D. J. Med. Chem. 1993, 36, 1090–1099; Song, Y.; Connor, D. T.; Sercel, A. D.; Sorenson, R. J.; Doubleday, R.; Unangst, P. C.; Roth, B. D.; Beylin, V. G.; Gilbertsen, R. B.; Chan, K.; Schrier, D. J.; Guglietta, A.; Bornemeier, D. A.; Dyer, R. D. J. Med. Chem. 1999, 42, 1161–1169; Labanauskas, L.; Kalcas, V.; Udrenaite, E.; Gaidelis, P.; Brukstus, A.; Dauksas, A. Pharmazie 2001, 56, 617–619; Boschelli, D. H.; Connor, D. T.; Bornemeier, D. A.; Dyer, R. D.; Kennedy, J. A.; Kuipers, P. J.; Okonkwo, G. C.; Schrier, D. J.; Wright, C. D. *J. Med. Chem.* 1993, 36, 1802–1810.

[45] Brezeanu, M.; Marinescu, D.; Badea, M.; Stanica, N.; Iles, M. A.; Supuran, C. T. Rev. Roum. Chim. 1997, 42, 727–732; Scozzafava, A.; Supuran, C. T. J. Enzyme Inhib. 1998, 13, 103–123; Supuran, C. T.; Clare, B. W. *Eur. J. Med. Chem.* 1999, 34, 41–50.

[46] Zareba, S. *Pharmazie* 1993, 48, 782–783.

[47] Chen, S. L.; Ji, S. X.; Zhu, Z. H.; Yao, Z. G. *Dyes Pigments* 1993, 23, 275–283.

[48] Choi, U. S.; Kim, T. W.; Jung, S. W.; Kim, C. J. *Bull. Korean Chem.* Soc. 1998, 19, 299–307.

[49] Gao, Y. J.; Zhang, Z. J.; Xue, Q. *J. Mater. Res. Bull.* 1999, 34, 1867–1874.

[50] Sun, X. W.; Hui, X. P.; Chu, C. H.; Zhang, Z. Y. Indian J. Chem., Sect. B 2001, 40, 15–19; Xu, P. F.; Yang, Y. P.; Wu, S. Z.; Zhang, Z. Y. Indian J. Chem., Sect. B 1998, 37, 127–131; Borg, S.; Estennebouhton, G.; Luthman, K.; Csoregh, I.; Hesselink, W.; Hacksell, U. J. Org. Chem. 1995, 60, 3112–3120; Mavrova, A. T.; Wesselinova, D.; Tsenov, Y. A.; Denkova, P. *Eur. J. Med. Chem.* 2009, 44, 63–69.

[51] Polshettiwar, V.; Varma, R. S. *Tetrahedron Lett.* 2008, 49, 879–883.

[52] Aryanasab, F.; Halimehjani, A. Z.; Saidi, M. R. *Tetrahedron Lett.* 2010, 51, 790-792.

[53] Al-Masoudi, I. A.; Al-Soud, Y. A.; Al-Salihi, N. J.; Al-Masoudi, N. A. Chem. Heterocycl. Compd. 2006, 42, 1377–1404. Haasnoot, J. G. *Coord. Chem. Rev.* 2000, 200–202, 131–185.

[54] Fischer, G. Adv. Heterocycl. Chem. 2008, 95, 143–219; Yang, G.; Yang, H. Trends Heterocycl. Chem. 2003, 9, 109–116; Fischer, G. Adv. Heterocycl. Chem. 1993, 57, 81–138; Fischer, G. Z. Chem. 1990, 30, 305–315. Dolzhenko, A. V.; Dolzhenko, A. V.; Chui, W. K. *Heterocycles* 2006, 68, 1723– 1759.

[55] Naito, Y.; Akahoshi, F.; Takeda, S.; Okada, T.; Kajii, M.; Nishimura, H.; Sugiura, M.; Fukaya, C.; Kagitani, Y. J. Med. Chem. 1996, 39, 3019–3029; Lipinski, C. A. *J. Med. Chem.* 1983, 26, 1–6;

[56] Dolzhenko, A. V.; Pastorin, G.; Dolzhenko, A. V.; Chui, W. K. *Tetrahedron Lett.* 2009, 50, 2124-2128.

[57] Bossert, F.; Meyer, H.; and Wehinger, E. Angew. *Chem., Int. Ed. Engl.,* 1981, 20, 762.

[58] Stout, D. M. and Meyers, A. I. Chem. Rev., 1982, 82, 223; Janis, R. A. and Triggle, D. J. *J. Med. Chem.,* 1983, 26, 775.

[59] Wang, G.W.; Miao C.B. *Green Chem.,* 2006, 8, 1080-85.

[60] Howard, A. S. In Comprehensive Heterocyclic Chemistry II; Katritzky, A. R., Rees, C. W., Scriven, E. V. F., Eds.; Pergamon Press: London, 1996; Vol. 8, pp 262–274, Chapter 10, and references therein. 4. Ismail, M. A.; Brun, R.; Wenzler, T.; Tanious, F. A.; Wilson, W. D.; Boykin, D. W. J. Med. Chem. 2004, 47, 3658–3664; Gudmundsson, K. S.; Williams, J. D.; Drach, J. C.; Townsend, L. B. J. Med. Chem. 2003, 46, 1449–1455; Geronikaki, A.; Babaev, E.; Dearden, J.; Dehaen, W.; Filimonov, D.; Galaeva, I.; Krajneva, V.; Lagunin, A.; Macaev, F.; Molodavkin, G.; Poroikov, V.; Pogrebnoi, S.; Saloutin, V.; Stepanchikova, A.; Stingaci, E.;

Tkach, N.; Vlad, L.; Voronina, T. *Bioorg. Med. Chem.* 2004, 12, 6559–6568.

[61] Adib, M.;Mahdavi, M.;Noghania, M.A.; Mirzaeib, P. *Tetrahedron Lett.* 2007, 48, 7263-7265.

[62] Tang, Y.; Oppenheimer, J.; Song, Z.; You, L.; Zhang, X.; and. Hsung, R. P.; *Tetrahedron,* 2006, 62, 10785; McKee, T. C.; Fuller, R. W.; Covington, C. D.; Cardellina, J. H.; Gulakowski, R. J.; Krepps, B. L.; McMahon, J. B and Boyd M. R.; *J. Nat. Prod.,* 1996, 59, 754. McKee, T. C.; Covington, C. D.; Fuller, R. W.; Bokesch, H. R Young, S.; Cardellina II, Kadushin, J. H.; Doel Soejarto, M. R.; Stevens, D.; Cragg, G. M and Boyd, M. R.; *J. Nat. Prod.,* 1998, 61, 1252; Grundon, M. F.; in *The Alkaloids: Quinoline Alkaloids Related to Anthranilic Acid,*Academic Press, London, 1988, vol. 32, p.341; Ulubelen, A.; Mericli, A. H.; Mericli F. and Kaya,;u. *Phytochemistry,* 1994, 35, 1600; Wu, S.-J. and Chen, I.-S.; *Phytochemistry,* 1993, 34, 1659; Campbell W. E.. Davidowitz, B and Jackson G. E.; *Phytochemistry,* 1990, 29, 1303. Khalid S.A and Waterman P.G., *Phytochemistry,* 1981, 20, 2761;

[63] Kumar, S. ; Hernandez, D.; Hoa, B.; Lee, Y.; Yang, J. S. and McCurdy, A. Org. Lett., 2008, 10, 3761; Rawat, M.; Prutyanov, V. and Wulff, W. D. *J. Am. Chem. Soc.,* 2006, 128, 11044; Fedorova, O.A.; Maure, F.; Chebunkova, A. V.; Strokach, Yu. P.; Valova, T. M.; Kuzmina, L. G. ; Howard, J. A. K.; Wenzel, M.; Gloe, K.; Lokshin, V. and Samat, A. *J. Phys. Org. Chem.,* 2007, 20, 469; Hepworth, J. D. and Heron, B. M. Prog. Heterocycl. Chem., 2005, 17, 33; Delbaer, S.; Micheau, J.-C. and Vermeersch, G. *J. Org. Chem.,* 2003, 68, 8968.

[64] Huber, C.; Moreau, J.; Batany, J.; Duboc, A.; Hurvois, J. P. Renaud, J. L.Adv. Synth. Catal., 2008, 350, 40; Moreau, J.; Hubert, C.; Batany, J.; Toupet, L.; Roisnel, T.; Hurvois, J.-P.; and Renaud, J.-L. *J. Org. Chem.,* 2009, 74, 8963.

[65] Garcia, A.; Borchardt, D.; Chang, C.-E. A.; Marsella, M. J. *J. Am.Chem. Soc.,* 2009, 131, 16640.

[66] Jung, E. J.; Park, B. H. and Lee, Y. R. *Green Chem.*2010, 12, 2003-2011.

[67] Kennedy, R. O. and Thornes, R. D. Coumarins: Biology, Applications and Mode of Action; Wiley: Chichester, UK, 1997; Murray, R. D. H.; Medez, J. and Brown, S. A. *The Natural Coumarins: Occurrence, Chemistry, and Biochemistry;* Wiley: New York, NY, 1982.

[68] Peter, H. *Dev. Oncol.* 1984, 22, 259.

[69] Dawood, K. M.; Ishii, H. and Fuchigami, T. J. Org. Chem. 2001, 66, 7030; Al- Nakib, T.; Bezjak, V.; Meegan, M. J. and Chandy, R. *Eur. J. Med.*

Chem. 1990, 25, 455; Al-Nakib, T.; Bezjak, V.; Rashid, S.; Fullam, B. and Meegan, M. J. *Eur. J. Med. Chem.* 1991, 26, 221; Ares, J. J.; Outt, P. E.; Randall, J. L.; Johnston, J.; Murray, P. D.; O'Brien, L. M.; Weisshaar, P. S. and Ems, B. L. *Bioorg. Med. Chem. Lett.* 1996, 6, 995; Wang, H. K.; Bastow, K. F.; Cosentino, L. M. and Lee, K. H. *J. Med. Chem.* 1996, 39, 1975; Gotoda, S.; Takahashi, N.; Nakagawa, H.; Murakami, M.; Takechi, T.; Komura, T.; Uchida, T. and Takagi, Y. *Pestic. Sci.* 1998, 52, 309.

[70] Moghaddam, F. M.; Kiamehr, M.; Khodabakhshi, M. R.; Mirjafary, Z.; Fathi, S. and Saeidian, H. *Tetrahedron.* 2010, 66, 8615-8622.

[71] Lambert, R. W.; Martin, J. A.; Merrett, J. H.; Parker, K. E. B.; Thomas, G. J. *PCT Int. Appl.* WO9706178, 1997.

[72] Hideo, T. Chem. Abstr. 1981, 95, 80922b, *Jpn. Tokkyo Koho JP* 56005480, 1981.

[73] Poupelin, J. P.; Ruf, G. S.; Blanpin, O. F.; Narcisse, G.; Ernouf, G. U.; Lacroix, R. *Eur. J. Med. Chem.* 1978, 13, 67.

[74] Menchen, S. M.; Benson, S. C.; Lam, J. Y. L.; Zhen, W.; Sun, D.; Rosenblum, B. B.; Khan, S. H.; Taing, M. US Patent 6583168, 2003; Banerjee, A.; Mukherjee, A. K. *Stain Technol.* 1981, 56, 83.

[75] Bekaert, A.; Andrieux, J.; Plat, M. Tetrahedron Lett. 1992, 33, 2805; Buehler, C. A.; Cooper, D. E.; Scrudder, E. O. *J. Org. Chem.* 1943, 8, 316.

[76] Papini, P.; Cimmarusti, R. Gazz. Chim. Ital. 1947, 77, 142. Sen, R. N.; Sarkar, N. J. Am. Chem. Soc. 1925, 47, 1079. Ota, K.; Kito, T. Bull. *Chem. Soc. Jpn.* 1976, 49, 1167.

[77] Dabiri, M.; Baghbanzadeh, M.; Nikcheh, M.S.; Arzroomchilar, E. *Bioorg. Med. Chem.* 2008, 18, 436-438.

[78] Skommer, J.; Wlodkowic, D.; Matto, M.; Eray, M.; Pelkonen, J. Leuk. Res. 2006, 30, 322–331.Yu, N.; Aramini, J. M.; Germann, M. W.; Huang, Z. Tetrahedron Lett. 2000, 41, 6993–6996. Bonsignore, L.; Loy, G.; Secci, D.; Calignano, A. Eur. J. Med. Chem. 1993, 28, 517–520. Witte, E. C.; Neubert, P.; Roesch, A. Ger. Offen. DE 3427985, 1986; Chem. Abstr. 1986, 104, 224915f. Andreani, L. L.; Lapi, E. *Boll. Chim. Farm.* 1960, 99, 583–587.

[79] Dabiri, M.; Bahramnejad, M.; Baghbanzadeh, M. *Tetrahedron,* 2009, 65, 9443–9447.

[80] Shaabani, A.; Samadi, S. and Rahmati, A. *Synthetic* Communications. 2007, 37, 491–499.

[81] Kappe, C. O. *Eur J Med Chem* 2000, 35, 1043.

[82] Singh, K.; Arora. D.; Singh, S. *Tetrahedron Lett.* 2006, 47, 4205

[83] Suzuki, I.; Suzumura, Y.; Takeda, K. *Tetrahedron Lett.* 2006, 47, 7861-7864.

[84] Mukhopadhyay, C.; Datta, A. J.Heterocyclic.Chem, 2010, 47, 136-146.

[85] Kidwai, M.; Saxena, S.; Khalilur Rahman Khan, M.; Thukral, S. S. *Eur. J. Med. Chem.* 2005, 40, 816.

[86] Hassani, Z.; Islami, M.R.; Kalantari, M. *Bioorg. Med. Chem.Lett.* 2006, 16, 4479-4482.

[87] Matsuoka, H.; Ohi, N.; Mihara, M.; Suzuki, H.; Miyamoto, K.; Maruyama, N.; Tsuji, K.; Kato, N.; Akimoto, T.; Takeda, Y.; Yano, K. and Kuroki, T.; *J. Med. Chem.*, 1997, 40, 105–111.

[88] Caliendo, G.; Grieco, P.; Perissutti, E.; Santagada, V.; Santini, A.; Alberizio, S.; Fattorusso, C.; Pinto, A.; and Sorrentino, R.; *Eur. J. Med. Chem.* 1998, 33, 957–967; Caliendo, G.; Perissutti, E.; Santagada, V.; Fiorino, F.; Severino, B.; Lippolis, L.; Pinto, A.; and Sorrentino, R. *Bioorg. Med. Chem.* 2002, 10, 2663–2669; Sebille, S.; de Tullio, P.; Boverie, S.; Antoine, M. H.; Lebrun, P.; and Pirotte, B.; *Curr. Med. Chem.* 2004, 11, 1213–1222.

[89] Yamashiro, D.; and Blake, J.; Int. J. Pept. Protein Res., 1981, 18, 383– 392; Mitin, Y. V.; and Zapevalova, N. P.; Int. J. Pept. Protein Res., 1990, 35, 352–356.

[90] Tefanic, P.; Anderluh, M.; Ilas, J.; Mravljak, J.; Sollner-Dolenc, M.; Stegnar M.; and Kikelj, D.J.; *Med. Chem.*, 2005, 48, 3110–3113; Stefanic, P.; Simoncic, Z.; Breznik, M.; Plavec, J.; Anderluh, M.; Addicks, E.; Giannis A.; and Kikelj, D.; *Org. Biomol. Chem.*, 2004,2, 1511–1517, Llas, J.; Anderluhc, P. S.; Dolenc, M. S.; and Kikelj, D.; *Tetrahedron,* 2005, 61, 7325–7348.

[91] Heravi, M. M.; Nami, N.; Oskooie, H. A. and Hekmatshoar, R. *Phosphorus, Sulfur, and Silicon, and the Related Elements.* 2005, 180, 1873-1878; Yavari, I.; Adib M.; and Hojabri, L.*Tetrahedron.* 2002, 58, 6895-6899; Touzeau, F.; Arrault, A.; Guillaumet, G.; Scalbert, E.; Pfeiffer, B.; Rettori, M.; Renard P. and Merour, J. Y.J. *Med. Chem.* 2003, 46, 1962-1979; Bourlot, A. S.; Sanchez, I.; Dureng, G.; Guillaumet, G.; Massingham, R.; Monteil, A.; Winslow, E.; Pujol M. D.; and Merour, J. Y. *J. Med. Chem.* 1998, 41, 3142-3158; Kundu, N. G.; Chaudhuri, G. and Upadhyay, A. J. Org. Chem. 2001, 66, 20-29; Yamazaki, A.; Achiwa, I and Achiwa, K. *Tetrahedron: Asymmetry.* 1996, 7, 403–406; Albanese, D.; Landini, D and Penso, M. *Chem. Commun.* 1999, 2095–2096; Buon, C.; Chacun-Lef′evre, L.; Rabot, R.; Bouyssou, P and Coudert, G. *Tetrahedron,* 2000, 56, 605-614; Torraca, K. E.; Kuwabe, S. I. and Buchwald, S. L. *J. Am. Chem. Soc.*

2000, 122, 12907-12908; Kuwabe, S.; Torraca, K. and Buchwald, S. L. *J. Am. Chem. Soc.* 2001, 123, 12202–12206; Cho, S.D.; Park, Y.D.; Kim, J. J.; Lee, S. G.; Ma, C.; Song, S. Y.; Joo, W. H.; Falck, J. R.; Shiro, M.; Shin D. S. and Yoon, Y. J.; *J. Org. Chem.* 2003, 68, 7918–7920.

[92] Caliendo, G.; Perissutti, E.; Santagada, V.; Fiorino, F.; Severino, B.; Cirillo, D.; dEmmanuele di Villa Bianca, R.; Lippolis, L.; Pinto, A. and Sorrentino, R. *Eur. J. Med. Chem.* 2004, 39, 815–826; Dai, W. M.; Wang, X and Ma, C. *Tetrahedron,* 2005, 61, 6879–6885; Feng, G. F.; Wu, J. L and Dai, W. M. *Tetrahedron,* 2006, 62, 4635– 4642.

[93] Kuroita, T.; Marubayashi, N.; Sano, M.; Kanzaki, K.; Inaba, K and Kawakita, T. *Chem. Pharm. Bull.* 1996, 44, 2051–2060; Caliendo, G.; Grieco, P.; Perissutti, E.; Santagada, V.; Santini, A.; Alberizio, S.; Fattorusso, C.; Pinto, A and Sorrentino R. *Eur. J. Med. Chem.*1998, 33, 957–967; Caliendo, G.; Perissutti, E.; Santagada, V.; Fiorino, F.; Severino, B.; dEmmanuele di Villa Bianca, R.; Lippolis, L.; Pinto, A and Sorrentino, R.; *Bioorg. Med. Chem.*, 2002, 10, 2663–2669; Combs, D. W.; Rampulla, M. S.; Bell, S. C.; Klaubert, D. H.; Tobia, A. J.; Falotico, R.; Haertlein, B.; Lakas-Weiss C and Moore, J. B.; *J. Med. Chem.,* 1990, 33, 380–386; Arrault, A.; Touzeau, F.; Guillaumet, G.; Leger, J. M.; Jarry, C and Merour, J. Y. *Tetrahedron,* 2002, 58, 8145– 8152.

[94] Zhang, Q.Y.; Liu, B.K.; Chen, W.Q.; Wu, Q.; and Lin, X.F. *Green Chem.* 2008, 10, 972-977.

[95] Cheeseman, G.W. H.; Cookson, R. F. In The Chemistry of Heterocyclic Compounds, 2nd ed.; Weissberger, A., Taylor, E. C., Eds.; Wiley: New York, NY, 1979; Vol. 35. pp 1-27. Porter, A. E. A. In *Comprehensive Heterocyclic Chemistry*; Katrizky, A. R., Rees, C. W., Eds.; Pergamon: New York, NY, 1984; Vol. 3, p 157.

[96] Dailey, S.; Feast, J. W.; Peace, R. J.; Sage, I. C.; Till, S.; Wood, E. L. *J. Mater. Chem.* 2001, 11, 2238; O'Brien, D.; Weaver, M. S.; Lidzey, D. G.; Bradley, D. D. C. *Appl. Phys. Lett.* 1996, 69, 881. Justin Thomas, K. R.; Marappan, V.; Jiann, T. L.; Chang Hao, C.; Yu-ai, T. *Chem. Mater.* 2005, 17, 1860. Brock, E. D.; Lewis, D. M.; Yousaf, T. I.; Harper, H. H. (The Procter & Gamble Company, USA) WO 9951688, 1999.

[97] Crossley, J. C.; Johnston, L. A. *Chem. Commun.* 2002, 1122.; Mizuno, T.; Wei, W. H.; Eller, L. R.; Sessler, J. L. *J. Am. Chem. Soc.* 2002, 124, 1134; Elwahy, A. H. M. *Tetrahedron* 2000, 56, 897.

[98] Lee, J.; Murray, W.V.; Rivero, R.A; J. Org. chem. 1997,62,3874; Zaragoza, F.; Stephensen, H.; *J. Org. Chem.,* 1999, 64, 2555.

[99] VOGEL's Textbook of Practical Organic Chemistry 5th ed., 1989, p 1190.; Brown, D. J. In *Chemistry of Heterocyclic Compounds, Quinoxalines Supplements II;* Taylor, E. C., Wipf, P., Eds.; John Wiley and Sons: New Jersey, 2004; Bhosale, R. S.; Sarda, S. R.; Andhapure, S. S.; Jadhav, W. N.; Bhusare, S. R.; Pawar, R. P. *Tetrahedron Lett.* 2005, 46, 7183; More, S. V.; Sastry, M. N. V.; Wang, C.; Yao, C. F. *Tetrahedron Lett.* 2005, 46, 6345.

[100] Raw, S. A.; Wilfred, C. D.; Taylor, R. J. K. *Org. Biomol. Chem.* 2004, 2, 788; Kim, S. Y.; Park, K. H.; Chung, Y. K. *Chem. Commun.* 2005, 1321; Robinson, R. S.; Taylor, R. J. K. Synlett 2005, 1003; Cho, C. S.; Renb, W. X.; Shim, S. C. *Tetrahedron Lett.* 2007, 48, 4665.

[101] Aparicio, D.; Attanasi, O. A.; Filippone, P.; Ignacio, R.; Lillini, S.; Mantellini, F.; Palacios, F.; Delos Santos, J. M. *J. Org. Chem.* 2006, 71, 5897.

[102] Das, B.; Venkateswarlu, K.; Suneel, K.; Majhi, A. *Tetrahedron Lett.* 2007, 48, 5371.

[103] Antoniotti, S.; Dunach, E. *Tetrahedron Lett.* 2002, 43, 3971.

[104] Liu, X. Y.; Che, C. M. *Angew. Chem. Int. Ed.* 2008, 47, 3805-3810.

[105] Murthy, S.N.; Madhav, B.; Nageswar, Y.V.D. *Helv. Chim. Acta.* 93, 2010, 1216-1220.

[106] Rao, K. S.; Lee, Y. R. *Tetrahedron.* 2010, 66, 8938-8944.

[107] Katritzky, A. R.; Rees, C. W., Eds.; Pergamon Press: Oxford, 1984; Vols. 1-8. Flitsch, W. In *Comprehensive Heterocyclic Chemistry* II; Katritzky, A. R., Rees, C. W., Scriven, E. F. V., Eds.; Pergamon: Oxford, U.K., 1996; Vol. 8, p 237.

[108] Micheal, J. P. Alkaloids 2001, 55, 91. Micheal, *J. P. Nat. Prod. Rep.* 2002, 19, 742

[109] Kel'in, A. V.; Sromek, A. W.; Gevorgyan, V. *J. Am. Chem. Soc.* 2001, 123, 2074. Kim, J. T.; Gevorgyan, V. Org. Lett. 2002, 4, 4697. Kim, J. T.; Butt, J.; Gevorgyan, V. *J. Org. Chem.* 2004, 69, 5638. Kim, J. T.; Gevorgyan, V. *J. Org. Chem.* 2005, 70, 2054. Seregin, I. V.; Gevorgyan, V. *J. Am. Chem. Soc.* 2006, 128, 12050. Smith, C. R.; Bunnelle, E. M.; Rhodes, A. J.; Sarpong, R. *Org. Lett.* 2007, 9, 1169.

[110] Liu, Y. H.; Song, Z. Q.; Yan, B. *Org. Lett.* 2007, *9*, 409.; Yan, B.; Zhou, Y. B.; Zhang, H.; Chen, J. J.; Liu, Y, H. *J. Org. Chem.*, Yan, B.; Liu, Y. *Org. Lett.* 2007, 9, 21, 4323-4326.

[111] Narayan, A. R. H.; Sarpong, R. *Green Chem.,* 2010, 12, 1556-1559.

[112] Guzikowski, A. P.; Whittemore, E. R.; Woodward, R. M.; Weber, E.; Keana, J. F. W. *J. Med. Chem.* 1997, 40, 2424–2429; Murakami, Y.; Hara, H.; Okada, T.; Hashizume, H.; Kii, M.; Ishihara, Y.; Ishikawa, M.;

Shimamura, M.; Mihara, S.; Kato, G.; Hanasaki, K.; Hagishita, S.; Fujimoto, M. *J. Med. Chem.* 1999, 42, 2621– 2632; Schultz, C.; Link, A.; Leost, M.; Zaharevitz, D. W.; Gussio, R.; Sausville, E. A.; Meijer, L.; Kunick, C. J. *Med. Chem.* 1999, 42, 2909–2919; Aramaki, Y.; Seto, M.; Okawa, T.; Oda, T.; Kanzaki, N.; Shiraishi, M. *Chem. Pharm. Bull.* 2004, 53, 254–258; Kunick, C.; Lauenroth, K.; Wieking, K.; Xie, X.; Schultz, C.; Gussio, R.; Zaharevitz, D.; Leost, M.; Meijer, L.; Weber, A.; Jorgensen, F. S.; Lemcke, T. *J. Med. Chem.* 2004, 47, 22–36. Bower, J. F.; Szeto, P.; Gallagher, T. *J. Org. Biomol. Chem.* 2007, 5, 143.

[113] Smalley, R. K.. In *Comprehensive Heterocyclic Chemistry;* Lwowski, W., Ed.; Pergamon Press: Bristol, 1984; 7, 545–546; Meigh, J. P. K.. In Science of Synthesis; Weinreb, S. M., Ed.; Thieme: Stuttgart, New York, 2004; 17, 829–830.

[114] Quadir, M.; Priestley, R. E.; Rising, T. W. D. F.; Gelbrich, T.; Coles, S. J.; Hursthouse, M. B.; Sheldrake, P. W.; Whittall, N.; Hii, K. K. *Tetrahedron Lett.* 2003, 44, 3675–3678. Dyker, G.; Markwitz, H. Synthesis 1998, 1750–1754. Martinez-Estibalez, U.; Sotomayor, N.; Lete, E. *Tetrahedron Lett.* 2007, 48, 2919– 2922. Suzuki, A.; Brown, H. C.. In *Organic Synthesis via Boranes; Aldrich Chemical Company*, Inc.: Milwaukee, USA, 2003; Vol. 3; for selected reviews on SM crosscoupling Miyaura, N.; Suzuki, A. *Chem. Rev.* 1995, 95, 2457–2483; Kotha, S.; Lahiri, K.; Kashinath, D. *Tetrahedron* 2002, 58, 9633–9695; Moreno- Manas, M.; Pleixats, R.; Sebastian, R. M.; Vallribera, A.; Roglans, A. *J. Organomet. Chem.* 2004, 689, 3669–3684; Nicolaou, K. C.; Bulger, P. G.; Sarlah, D. *Angew. Chem.* 2005, 117, 4516–4563; *Angew. Chem.,* Int. Ed. 2005, 44, 4442–4489.; Kotha, S.; Lahiri, K. *Eur. J. Org. Chem.* 2007, 1221–1236; Kotha, S.; Chakraborty, K.; Brahmachary, E. Synlett 1999, 1621–1623. Trybulski, E. J.; Reeder, E.; Blount, J. F.; Walser, A.; Fryer, R. I. *J. Org. Chem.* 1982, 47, 2441–2447; Jenny-Lee, P.; Pathak, R.; De Koning, C. B.; Van Otterlo, W. A. L. *Eur. J. Org. Chem.* 2007, 4953–4961.

[115] Prasad, J. V.; Prabhakar, M.; Manjulatha, K.; Rambabu, D.; Anand Solomon, K.; Krishna, G. G.; Kumar, K.A. *Tetrahedron Lett.* 2010, 51, 3109-3111.

[116] Schutz, H. Benzodiazepines, Springer, Heidelberg, 1982; Landquist, J. K. in *Comprehensive Heterocyclic Chemistry,* ed. Katritzky, A. R. Rees, C. W. Pergamon, Oxford, 1984, vol. 1,p. 166; L. O. Randall and B. Kamel, in Benzodiazepines,ed. Garattini, S. Mussini. E. Randall, L. O. Raven Press, New York, 1973, p. 27.

[117] Schreiber, S. L. Science, 2000, 287, 1964. ; Hester, J. B. Jr.; Rudzik A. D.; Kamdar, B. V.*J. Med. Chem.,* 1971, 14, 1078.

[118] Palimkar, S. S.; Lahoti, R. J.; Srinivasan, K. V. *Green Chem.,* 2007, 9, 146-152.

In: Water Condensation
Editor: Amimul Ahsan

ISBN: 978-1-61324-874-4

Chapter 4

THEORETICAL APPROACH OF CONDENSATION FOR SOLAR STILL

Amimul Ahsan[1,*], Monzur A. Imteaz[2], Md. Zahangir Alam[3], A. A. Mokammel Haque[4] and Abdul Halim Ghazali[1]

[1]University Putra Malaysia, Department of Civil Engineering, Faculty of Engineering, UPM Serdang, Selangor, Malaysia
[2]Swinburne University of Technology, Faculty of Engineering and Industrial Science, Hawthorn, Melbourne, Australia
[3]International Islamic University Malaysia, Department of Biotechnology Engineering, Faculty of Engineering, Gombak, Kuala Lumpur, Malaysia
[4]Universiti Sains Malaysia, School of Civil Engineering, Engineering Campus, Nibong Tebal, Seberang Perai Selatan, P. Pinang, Malaysia

ABSTRACT

Solar water distillation might be one of the viable options for providing drinking water for a single house or a small community in an arid, remote or

[*]E-mail: ashikcivil@yahoo.com, aahsan@eng.upm.edu.my (A. Ahsan) Tel: 60389464492; Fax: 60386567129.

coastal region. A new type of solar distillation called a Tubular Solar Still (TSS) was designed to overcome some disadvantages in the maintenance and management of a basin type still. A production model of a TSS based on a film-wise condensation theory that takes into account of thermal resistance in unsaturated humid air inside the still is presented in this chapter. The condensation coefficient due to thermal resistance is inversely proportional to the dry air pressure fraction. In the present model, the overall heat transfer coefficient between the humid air inside the still and the ambient air outside the still is used for the first time because the measurement of the ambient air temperature is easier to obtain than the inner surface temperature of the tubular cover. The analytical solution of this condensation theory could provide a good agreement with the observed production flux obtained from the field experiment and proves that the model can be used to precisely predict the production flux. A basic theoretical approach on laminar film condensation is presented first in this chapter to facilitate readers understanding before introducing the theory of film-wise condensation for a solar still.

NOMENCLATURE

D	diameter of TSS (m);
e_a	partial dry air pressure (Pa);
e_o	total atmospheric pressure (101325 Pa);
e_{vha}	partial water vapor pressure in humid air (Pa);
g	gravitational acceleration (9.807 m/s^2);
h	heat transfer coefficient (W/m^2·K);
h_{fg}	latent heat of vaporization (J/kg);
$\dot{m}$	condensate mass flow per unit length (kg/m·s);
m	condensation mass flux (kg/m^2·s);
p_h	hourly production mass flux (kg/m^2·hr);
$\dot{q}$	heat flow per unit length (W/m);
q_s	substantial heat flux (W/m^2);
$\dot{q}_s$	apparent heat flux (W/m^2);
R	radius of TSS (m);
R_s	radiative heat flux (W/m^2);
RH	relative humidity (%);
s	wetted perimeter (m);

T	temperature (K);
T'	temperature of liquid film (K);
v_θ	angular velocity (m/s);
γ_c	condensation coefficient (-);
γ_1	temperature correction coefficient (-);
γ_2	area fraction (-);
δ	condensate film thickness (m);
θ	azimuthal angle (radians);
λ	thermal conductivity (W/m·K);
μ	dynamic viscosity (kg/m·s);
ρ	density (kg/m^3);
τ	shear stress (Pa);

Subscripts

a	ambient air;
c	condensation, condensate;
ci	inner surface of tubular cover;
h	hourly;
ha	humid air;
$i\theta$	interfacial shear stress in θ direction;
l	condensate liquid, local;
o	overall between humid and ambient air;
vha	water vapor in humid air; and
w	water surface.

1. Introduction

A solar still can be easily built on-site without using special tools to produce potable water from saline water using solar energy. This easy assembly helps to shorten the transportation distance of the water from the still to the point of use. A basin type still is the most common among solar stills, but is difficult to quickly repair and also to construct. A new type of solar distillation called a Tubular Solar Still (TSS) was designed [1-3] to overcome some disadvantages of the basin type still. The TSS consists of a transparent tubular cover made of vinyl chloride and a trough inside the cover. Consequently, the weight of the TSS becomes much lighter than that of the basin type with a typical glass cover.

Some production models of the basin type solar still have been proposed [4-7] using evaporative heat and mass transfer correlations to predict the distilled water output, i.e. production. The distilled water is, however, produced after the condensation which takes place on the inner surface of the tubular cover, following the evaporation from the water surface in a trough. Therefore, there exists a time lag between the beginning of the production of distilled water and the occurrence of the evaporation [2]. By observing the time lag in the evaporation $\rightarrow$ condensation $\rightarrow$ production process in the still, it can be deduced that the theory of condensation may give a better result for predicting the hourly production than the past evaporation theories [1].

No comprehensive condensation theory has yet been developed that clearly explains the heat and mass transfer phenomena due to condensation inside a solar still. Although some condensation theories [8-10] have been applied to a film flow achieved on a plate or a tube that derive the condensate mass flow and the heat transfer rate. Therefore, the theoretical approach of Nusselt [11] will be described first in this chapter to facilitate readers understanding on laminar film condensation before introducing a theory of condensation for the solar still. This chapter aims to introduce a film-wise condensation theory for the TSS that takes into account the thermal resistance of unsaturated humid air. The field experimental results of hourly production will be then compared with the theoretical predictions.

2. Condensation on a Vertical Plate

In 1916, Nusselt first derived a famous equation for the laminar film condensation heat transfer on the outside surface of a cooled vertical flat plate and a vertical cylindrical tube of inner diameter larger than 0.0032m [11]. The laminar film condensation on a vertical plate can be expressed as follows:

2.1. X-momentum Equation

X-momentum equation can be expressed as

$$U\frac{\partial u}{\partial x}+V\frac{\partial u}{\partial y}=-\frac{1}{\rho}\frac{dp}{dx}-g+\nu\frac{d^2u}{\partial y^2}$$

$$\frac{dp}{dx}=\left(\frac{dp}{dx}\right)_{y=\delta}=\frac{dp_\delta}{dx}$$

$$(1)$$

$$y=\delta \quad U=V=0 \quad \rho=\rho$$

$$0=-\frac{1}{\rho}\frac{dp}{dx}-g$$

$$\therefore \frac{dp_\delta}{dx}=-\rho_v g$$

There may be several complicating features associated with film condensation as shown in Figure 1. The film originates at the top of the plate and flows downward under the influence of gravity. The condensate film thickness δ and the condensate mass flow rate $\dot{m}$ increase with increasing x because of continuous condensation at the liquid-vapor interface, which is at T_{sat}.

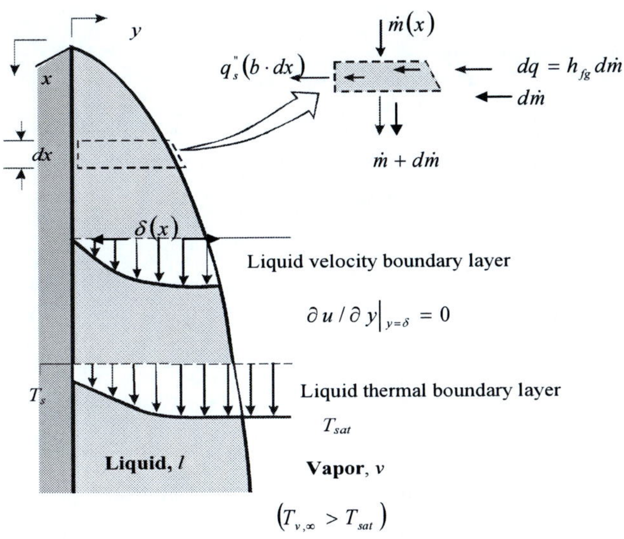

Figure 1. Nusselt's analysis for a vertical plate [11-12].

2.2. Assumptions

Despite the complexities associated with film condensation, useful results may be obtained by making some assumptions that originated with an analysis by Nusselt [11-12].

1. Laminar flow and constant properties are assumed for the liquid film.
2. The gas is assumed to be a pure vapor and at a uniform temperature equal to T_{sat}. With no temperature gradient in the vapor, heat transfer to the liquid-vapor interface can occur only by condensation at the interface and not by conduction from the vapor.
3. The shear stress at the liquid–vapor interface is assumed to be negligible, in which case $\partial u / \partial y\big|_{y=\delta} = 0$. With this assumption and the foregoing assumption of a uniform vapor temperature, there is no need to consider the vapor velocity or thermal boundary layers shown in Figure 1.
4. Momentum and energy transfer by advection in the condensate film are assumed to be negligible. This assumption is reasonable by virtue of the low velocities associated with the film. It follows that heat transfer across the film occurs only by conduction, in which case the liquid temperature distribution is linear.

2.3. Theoretical Analysis

Film conditions resulting from the assumptions are shown in Figure 1. From the forth approximation, momentum advection terms may be neglected, and the *x*-momentum equation may be expressed as [11-12]

$$\frac{\partial^2 u}{\partial y^2} = \frac{1}{\mu_l}\frac{d p}{d x} - \frac{X}{\mu_l} \tag{2}$$

Where the body force X has been retained. The body force within the film is equal to $\rho_l g$, and the pressure gradient may be approximated in terms of conditions outside the film. That is, invoking the boundary layer approximation that $(\partial p / \partial y) \approx 0$, it follows that $(d p / d x) \approx \rho_v g$. The momentum equation may therefore be expressed as [11-12]

$$\frac{\partial^2 u}{\partial y^2} = -\frac{g}{\mu_l}(\rho_l - \rho_v) \tag{3}$$

Integrating twice and applying boundary conditions of the from $u(0) = 0$ and $\partial u / \partial y|_{y=\delta} = 0$, the velocity profile in the film becomes

$$u(y) = \frac{g(\rho_l - \rho_v)\delta^2}{\mu_l}\left[\frac{y}{\delta} - \frac{1}{2}\left(\frac{y}{\delta}\right)^2\right] \tag{4}$$

From this result the condensate mass flow rate per unit width $\Gamma(x)$ may be obtained in terms of an integral involving the velocity profile

$$\frac{\dot{m}(x)}{b} = \int_0^{\delta(x)} \rho_l u(y)dy \equiv \Gamma(x) \tag{5}$$

Substituting from Eq. 4, it follows that

$$\Gamma(x) = \frac{g\rho_l(\rho_l - \rho_v)\delta^3}{3\mu_l} \tag{6}$$

The specific variation with *x* of δ, and hence of Γ, may be obtained by first applying the conservation of energy requirement to the differential element shown in Figure 1. At a portion of the liquid-vapor interface of unit width and length *dx*, the rate of heat transfer into the film, *dq*, must equal the rate of energy release due to condensation at the interface. Hence

$$dq = h_{fg}\, d\dot{m} \tag{7}$$

Since advection effects are neglected, it also follows that the rate of heat transfer across the interface must equal the rate of heat to the surface.

Hence

$$dq = q_s^{''}(b \cdot dx) \tag{8}$$

Since the liquid temperature distribution is linear, Fourier's law may be used to express the surface heat flux as

$$q_s^{''} = \frac{k_l(T_{sat} - T_s)}{\delta} \tag{9}$$

Combining Eqs. 5 and 7 to 9, we then obtain

$$\frac{d\Gamma}{dx} = \frac{k_l(T_{sat} - T_s)}{\delta h_{fg}} \tag{10}$$

Differentiating Eq. 6 we also obtain

$$\frac{d\Gamma}{dx} = \frac{g\rho_l(\rho_l - \rho_v)\delta^2}{\mu_l}\frac{d\delta}{dx} \tag{11}$$

Combining Eqs. 10 and 11, in follows that

$$\delta^3 d\delta = \frac{k_l \mu_l(T_{sat} - T_s)}{g\rho_l(\rho_l - \rho_0)h_{fg}}dx \tag{12}$$

Integrating from $x = 0$, where $\delta = 0$, to any x location of interest on the surface [11-12]

$$\delta(x) = \left[\frac{4k_l u_l(T_{sat} - T_s)x}{g\rho_l(\rho_l - \rho_v)h_{fg}}\right]^{1/4} \tag{13}$$

This result may then be substituted into Eq. 6 to obtain $\Gamma(x)$.

An improvement to the foregoing result for $\delta(x)$ was made by Nusselt [11] and Rohsenow [13], who showed that, with the inclusion of thermal advection effects, a term is added to the latent heat of vaporization. In lieu of h_{fg}, Rohsenow [13] recommended using a modified latent heat of the form $h'_{fg} = h_{fg} + 0.68c_{p,l}(T_{ssat} - T_s)$, or in terms of the Jakob number,

$$h'_{fg} = h_{fg}(1 + 0.68Ja) \tag{14}$$

Sadasivan and Lienhard [14] have shown that the modified latent heat depends weakly on the Prandtl number of the liquid.

The surface heat flux may be expressed as

$$q''_s = h_x(T_{sat} - T_s) \tag{15}$$

Substituting from Eq. 9, the local convection coefficient is then

$$k_x = \frac{k_l}{\delta} \tag{16}$$

or, from Eq. 13, with h_{fg} replaced by h'_{fg},

$$h_x = \left[\frac{g\rho_l(\rho_l - \rho_v)k_l^3 h'_{fg}}{4\mu_l(T_{sat} - T_s)x}\right]^{1/4} \tag{17}$$

Since h_x depends on $x^{-1/4}$, it follows that the average convection coefficient for the entire plate is

$$\bar{h}_L = \frac{1}{L}\int_0^L h_x dx = \frac{4}{3}h_L \tag{18}$$

or

$$\bar{h}_L = 0.943\left[\frac{g\rho_l(\rho_l - \rho_v)k_l^3 h'_{fg}}{\mu_l(T_{sat} - T_s)x}\right]^{1/4} \tag{19}$$

The average Nusselt number then has the form

$$Nu_L = \frac{\bar{h}_L L}{k_l} = 0.943\left[\frac{\rho_l g(\rho_l - \rho_v)h'_{fg} L^3}{\mu_l k_l(T_{sat} - T_s)}\right]^{1/4} \tag{20}$$

In using this equation all liquid properties should be evaluated at the film temperature $T_f = (T_{sat} - T_s)/2$, and h_{fg} should be evaluated at T_{sat}.

A more detailed boundary layer analysis of film condensation on a vertical plate has been performed by Sparrow and Gregg [15]. Their results confirmed by Chen [16], indicate that errors associated with using Eq. 20 are less than 3% for $Ja \leq 0.1$ and $1 \leq \Pr \leq 100$. Dhir and Lienhard [17] have also shown that Eq. 20 may be used for inclined plates, if g is replaced by $g \cdot \cos\theta$, where θ is the angle between the vertical and the surface. However, it must be used with caution for large values of θ and does not apply if $\theta = \pi / 2$. The expression may be used for condensation on the inner or outer surface of a vertical tube of radius R, if $R > \delta$.

The total heat transfer to the surface may be obtained by using Newton's law of cooling [12]

$$q = \vec{h}_L A(T_{sat} - T_s) \tag{21}$$

The total condensation rate may then be determined from the relation [12]

$$\dot{m} = \frac{q}{h'_{fg}} = \frac{\bar{h}_L A(T_{sat} - T_s)}{h'_{fg}} \tag{22}$$

Eqs. 21 and 22 are generally applicable to any surface geometry, although the form of $\bar{h}_L$ will vary according to geometry and flow conditions [12].

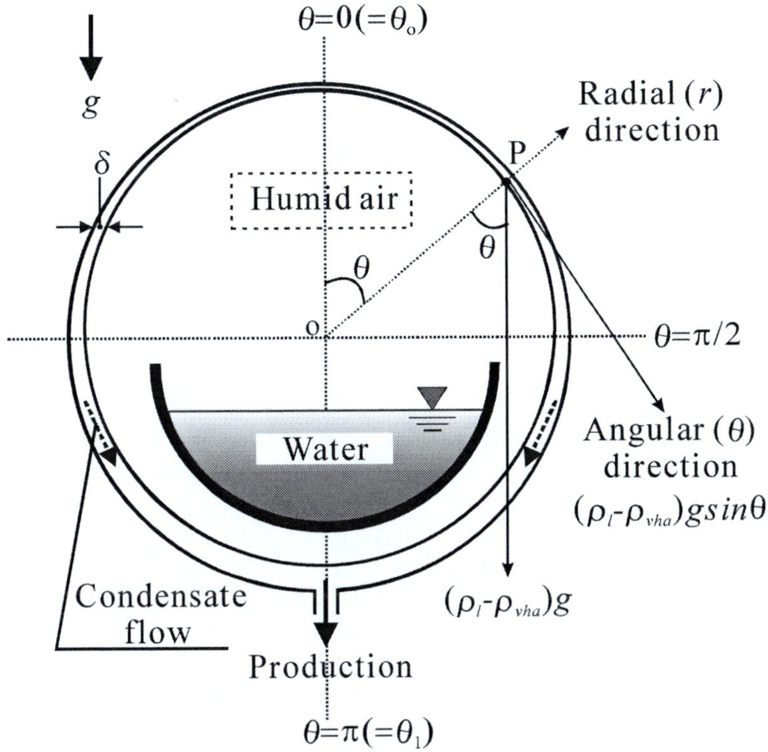

Figure 2. Condensation on inner surface of tubular cover [18-19].

3. Condensation inside a Tubular Still

The film-wise condensation theory [18-19] that takes into account of thermal resistance in unsaturated humid air inside the TSS is presented below.

3.1. Assumptions

The following assumptions are made to simplify the analysis [18-19]:

1. The TSS is cylindrical and set horizontally.
2. The condensate liquid film (liquid film) is a laminar flow and flows only along the inner circumference of the tubular cover, i.e. in the angular (θ) direction shown in Figure 2.

3. The liquid film is a single homogeneous flow in the axial (x) direction shown in Figure 3.
4. The temperatures of the humid air, T_{ha}, and the tubular cover, T_{ci}, are uniform.
5. The heat transfer at the liquid film-humid air interface (interface) is caused only by condensation. The convection in the liquid film associated with the angular velocity, v_θ does not contribute to the heat transfer in the liquid film.
6. The boundary layer of T_{ha} exists on the interface and the boundary layer thickness, δ_{ha}, or the surface temperature of the liquid film, T_{ha}', might be affected by the thermal resistance associated with the presence of a non-condensable gas (dry air) in the humid air (Figure 4(a)).
7. The interfacial shear stress at the interface is negligible, i.e. $\tau_{i\theta}=0$.
8. The velocity (v_θ) gradient in the θ direction is negligible compared to the radial (r) direction.

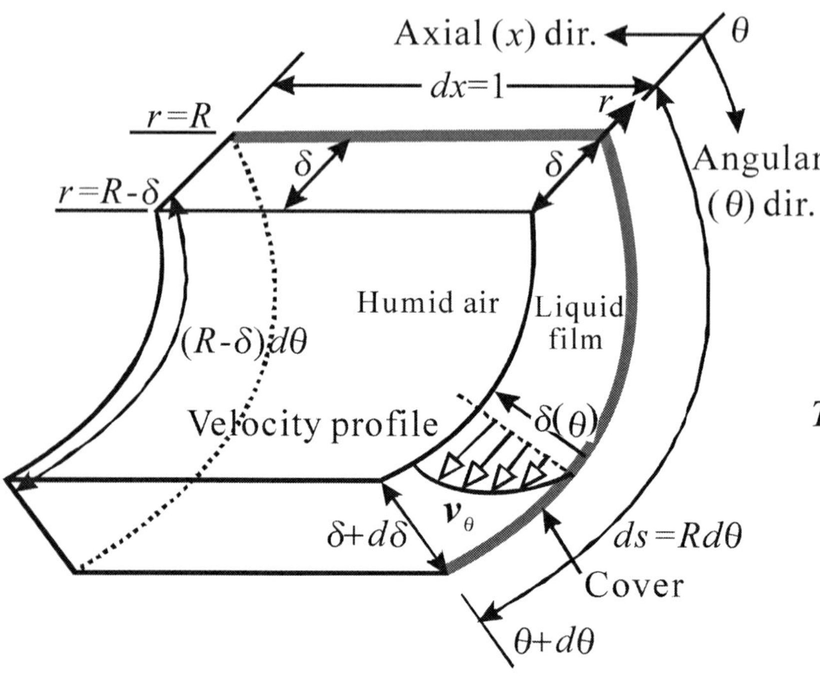

Figure 3. Representative element of condensate flow [18-19].

3.2. Momentum and Energy Equations

Then the governing equations of momentum and energy for a liquid film under a steady state condition can be simplified as follows [18-19]:

Momentum equation:

$$\left(\frac{d^2 v_\theta}{dr^2}+\frac{1}{r}\frac{dv_\theta}{dr}-\frac{v_\theta}{r^2}\right)+\frac{g}{\mu_l}(\rho_l-\rho_{vha})\sin\theta=0 \tag{23}$$

Energy equation:

$$\frac{d^2 T}{dr^2}+\frac{1}{r}\frac{dT}{dr}=0 \tag{24}$$

The governing equations are subjected to the following boundary conditions:

$$v_\theta=0;\ T=T_{ci}\ at\ r=R, 0\le\theta\le\pi \tag{25}$$

$$\tau_{i\theta}=\mu_l\left[r\frac{d}{dr}\left(\frac{v_\theta}{r}\right)\right]=0;\ T=T_{ha}'\ at\ r=R-\delta(\theta), 0\le\theta\le\pi \tag{26}$$

where, g = gravitational acceleration; μ_l = dynamic viscosity of the condensate liquid; ρ_l = density of the condensate liquid; ρ_{vha} = density of water vapor in the humid air; R = radius of the TSS; and δ = condensate film thickness. When integrating Eq. 23 is subjected to the boundary conditions of Eqs. 25 and 26, the profile of v_θ in the r direction becomes

$$v_\theta=\left[\frac{XR}{3}+\frac{X(R-\delta)^3}{6R^2}\right]r-\frac{X(R-\delta)^3}{6r}-\frac{Xr^2}{3} \tag{27}$$

where, $X=\frac{g}{\mu_l}(\rho_l-\rho_{vha})\sin\theta$.

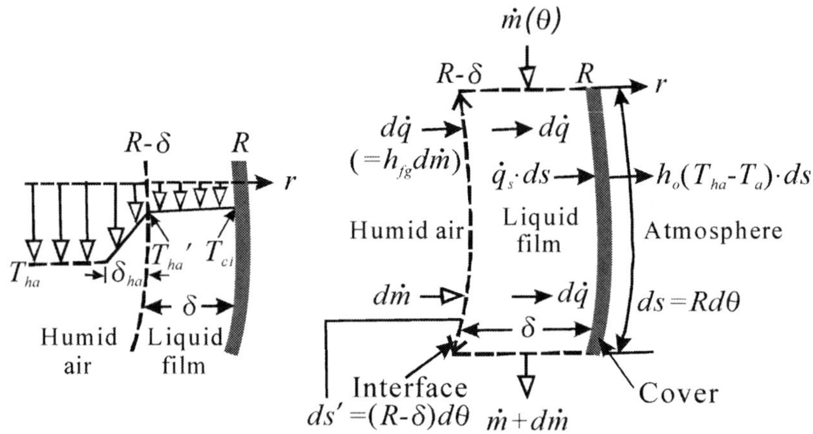

(a) Temperature profile (b) Heat and mass transfer

Figure 4. Temperature profile and heat and mass transfer in condensate flow [18-19].

Using Fourier's law, the substantial heat flux for the neat area of the liquid film, q_s, at the interface ($r=R-\delta$) is given by

$$q_s = -\lambda_l \left.\frac{dT}{dr}\right|_{r=R-\delta} \tag{28}$$

where, λ_l = thermal conductivity of the condensate liquid.

Integrating Eq. 24 subjected to the boundary conditions, $T = T_{ci}\big|_{r=R}$ and $T = T_{ha}'\big|_{r=R-\delta}$, yields

$$\left.\frac{dT}{dr}\right|_{r=R-\delta} = \frac{T_{ha}' - T_{ci}}{(R-\delta)\ln\left(\frac{R-\delta}{R}\right)} \tag{29}$$

Substituting Eq. 29 into Eq. 28, q_s is given by

$$q_s = \frac{\gamma_1 \lambda_l (T_{ha} - T_{ci})}{(R-\delta)\ln\left(\frac{R}{R-\delta}\right)} \tag{30}$$

where, $\gamma_1 = \frac{T_{ha}' - T_{ci}}{T_{ha} - T_{ci}}$ (<1). γ_1 is called the temperature correction coefficient, which might be affected by the thermal resistance at the interface.

Since the liquid film thickness is fairly small compared to the radius of the TSS (i.e. $\delta << R$), Eq. 30 is approximated as

$$q_s = \frac{\gamma_1 \lambda_l (T_{ha} - T_{ci})}{\delta(\theta)} \tag{31}$$

The apparent heat flux for the whole area of the tubular cover, $\dot{q}_s$, can be expressed using Eq. 31 as

$$\dot{q}_s = q_s \gamma_2 = \frac{\gamma_c \lambda_l (T_{ha} - T_{ci})}{\delta(\theta)} \tag{32}$$

where, γ_2 (<1) is the area fraction (=net area of the liquid film/whole area of the tubular cover) and might be influenced by the wettability of the cover material. γ_c (=$\gamma_1\gamma_2$) is called the condensation coefficient.

According to the assumption 5., the heat flow across the interface per unit length (dx=1) due to condensation is equal to that across the inner surface of the tubular cover shown in Figure 4(b). That is,

$$d\dot{q} = h_{fg} d\dot{m} = \dot{q}_s \cdot ds \tag{33}$$

where, $\dot{q}$ = heat flow per unit length; h_{fg} = latent heat of vaporization; $\dot{m}$ = condensate mass flow per unit length; and s = wetted perimeter. Using $ds=Rd\theta$ and Eq. 33, $\dot{q}_s$ is written as [18-19]

$$\dot{q}_s = h_{fg} \frac{d\dot{m}}{Rd\theta} = h_{fg} \frac{d\dot{m}}{d\delta} \frac{d\delta}{Rd\theta} \tag{34}$$

3.3. Film Thickness

The variation of δ in the θ direction can be expressed as [18-19]

$$\delta(\theta) = \left[4A \ln \left| \frac{\tan\theta/2}{\tan\theta_o/2} \right| \right]^{1/4} \tag{35}$$

where, $A = \dfrac{\gamma_c \lambda_l \mu_l D(T_{ha} - T_{ci})}{2g\rho_l(\rho_l - \rho_{vha})h_{fg}}$; and D=2R.

3.4. Heat Transfer Coefficient

The average heat transfer coefficient, h_c, is given as [18-19]

$$h_c = 0.996\gamma_c^{\frac{3}{4}} \left[\frac{g\rho_l(\rho_l - \rho_{vha})h_{fg}\lambda_l^3}{\mu_l D(T_{ha} - T_{ci})} \right]^{1/4} \tag{36}$$

Supposing that the temperature difference fraction, $a=(T_{ha}-T_{ci})/(T_{ha}-T_a)$, is constant, the overall heat transfer coefficient, h_o, is expressed as [18-19]

$$h_o = 0.996\gamma_c^{\frac{3}{4}} \left[\frac{g\rho_l(\rho_l - \rho_{vha})h_{fg}a^3\lambda_l^3}{\mu_l D(T_{ha} - T_a)} \right]^{1/4} \tag{37}$$

Thus, the condensate mass flux, m, is defined as

$$m = h_o(T_{ha} - T_a)/h_{fg} \tag{38}$$

Substituting Eq. 37 into Eq. 38, γ_c is given by

$$\gamma_c = \left(\frac{m}{0.996}\right)^{\frac{4}{3}} \left[\frac{\mu_l D h_{fg}^3}{g\rho_l(\rho_l - \rho_{vha})a^3\lambda_l^3(T_{ha} - T_a)^3}\right]^{\frac{1}{3}} \quad (39)$$

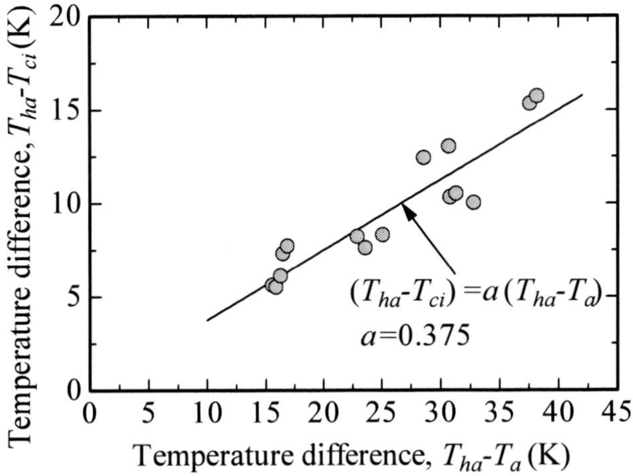

Figure 5. Relation between temperature difference, T_{ha}-T_{ci}, and temperature difference, T_{ha}-T_a [2, 18-19].

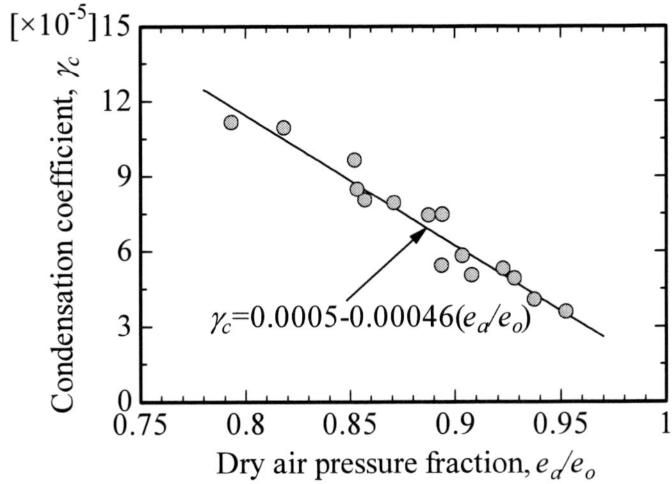

Figure 6. Relation between condensation coefficient, γ_c, and dry air pressure fraction, e_a/e_o [2, 18-19].

3.5. Condensation Coefficient

Figure 5 shows the relation between T_{ha}-T_{ci} and T_{ha}-T_a. As can be seen from the straight regression line, the value of a in Eq. 39 is 0.375. Since the condensation rate was not measured, γ_c was calculated by Eq. 39 under the assumption that the condensation flux is equal to the production flux [18-19].

Since the humid air is definitely not saturated, it is inferred that γ_c would be affected by the thermal resistance due to the presence of a dry air. Based on this concept, Figure 6 shows the relation between γ_c and the dry air pressure fraction, e_a/e_o (=partial dry air pressure/total atmospheric pressure). The value of γ_c is inversely proportional to e_a/e_o. Thus, h_o is expressed by [18-19]

$$h_o = \left(1.86 - 1.72\frac{e_a}{e_o}\right)^{\frac{3}{4}} \left[\frac{g\rho_l(\rho_l - \rho_{vha})h_{fg}\lambda_l^{\,3}}{\mu_l D(T_{ha} - T_a)}\right]^{\frac{1}{4}} \times 10^{-3} \tag{40}$$

where, $\frac{e_a}{e_o} = 1 - \frac{e_{vha}}{e_o}$; $e_{vha} = f(T_{ha}, RH_{ha})$.

The hourly condensation mass flux, m_h, is defined as

$$m_h = \frac{3600\, h_o(T_{ha} - T_a)}{h_{fg}} \tag{41}$$

3.6. Condensation flux

Figure 7 shows the diurnal variations of p_h obtained from the field experiment and of m_h calculated by Eq. 41. The profile of the p_h is almost the same as that of the m_h. The time lag between the condensation and the production is negligible as far as the TSS used in the experiment is concerned [18-19].

Figure 8 shows the comparison between the p_h and the m_h. Both correspond well with each other. From Figures 7 and 8, it is revealed that the p_h can be substituted for m_h [18-19]. Readers are referred to references [20-21] for further experimental study and numerical analysis on solar still.

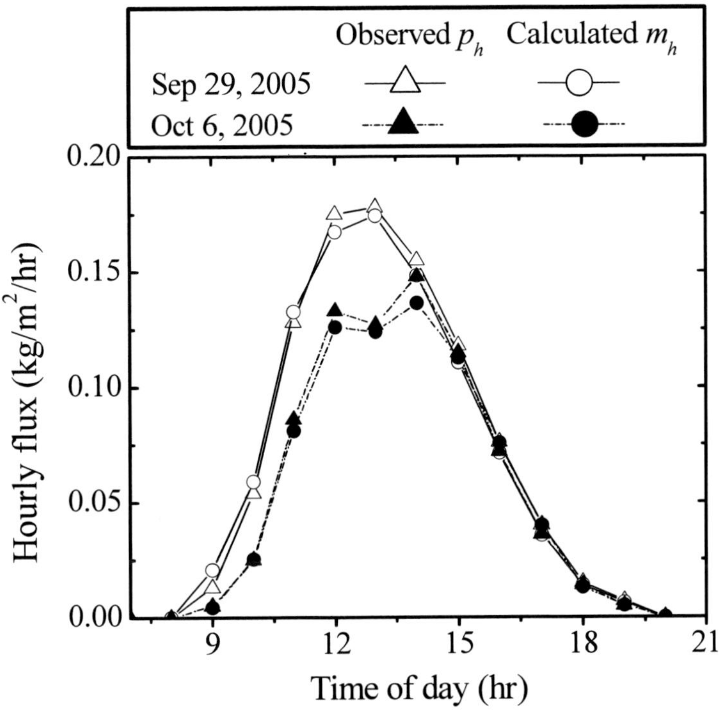

Figure 7. Diurnal variations of observed production flux, p_h, and calculated condensation flux, m_h [2, 18-19].

CONCLUSION

A film-wise condensation theory that takes into account of thermal resistance in unsaturated humid air inside a Tubular Solar Still was developed in this chapter. It was revealed that the condensation coefficient was in reverse proportion to the dry air pressure fraction. The present condensation theory using an overall heat transfer coefficient between humid and ambient air. The validity of the condensation theory was evaluated with a comparison between the field experimental data and the theoretical results. Due to the correspondence of these results it can be concluded that the present condensation theory can be used to precisely predict the production flux of a solar still.

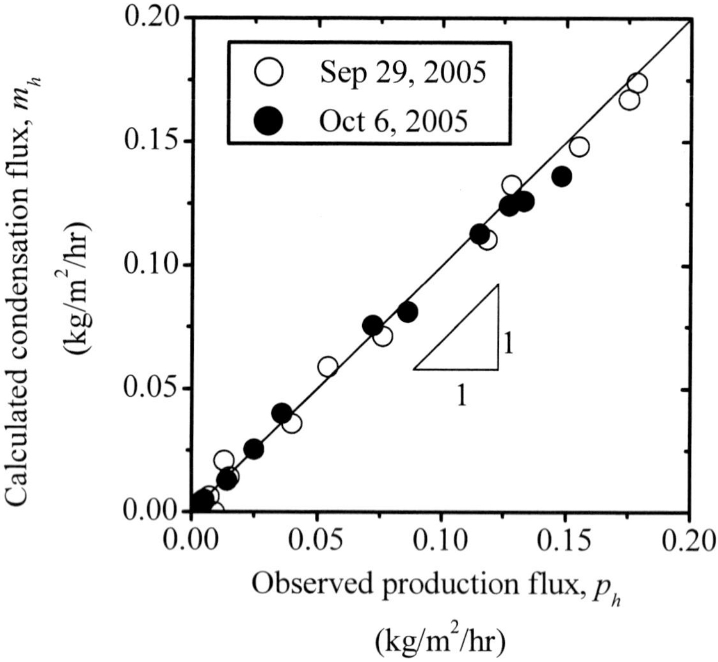

Figure 8. Comparison of calculated condensation flux, m_h, with observed production flux, p_h [2, 18-19].

Acknowledgments

Authors gratefully acknowledge Prof. Dr. Teruyuki Fukuhara, Engr. Keiichi Waki, Dr. Shafiul Islam, Engr. Hiroaki Terasaki, Dr. Akihiro Fujimoto, Dr. Yasuo Kita, Dr. Kazuo Okamura and Engr. Fumio Asano for their kind cooperation during staying in Fukui, Japan and for their continued friendly support. The partial financial support provided by the Ministry of Education, Science, and Culture of Japanese Government, Japan; Shimizu Corporation, Japan and Japan Cooperation Center, Petroleum (JCCP), Japan is also acknowledged.

References

[1] Ahsan, A. *Production model of new tubular solar still and its productivity characteristics*. PhD thesis, University of Fukui, Japan, 2009, pp 45-73.

[2] Islam, K. M. S. *Heat and vapor transfer in tubular solar still and its production performance*. PhD thesis, University of Fukui, Japan, 2006, pp 33-52.

[3] Fukuhara, T.; Islam, K. M. S. Tubular solar desalination and improvement of soil moisture retention by date palm. In *Arid Land Hydrogeology: In Search of a Solution to a Threatened Resource*. Mohamed, A. M. O.; Ed.; Taylor and Francis, London, 2006; pp 153-162.

[4] Chaibi, M. T. Analysis by simulation of a solar still integrated in a greenhouse roof, *Desalination*, 2000, vol. 128, 123-138.

[5] Clark, J. A. The steady-state performance of a solar still, *Solar Energy*, 1990, vol. 44 (1), 43-49.

[6] Hongfei, Z.; Xiaoyan, Z.; Jing, Z.; Yuyuan, W. A group of improved heat and mass transfer correlations in solar stills, *Energy Con. Mang.*, 2002, vol. 43, 2469-2478.

[7] Shawaqfeh, A. T.; Farid, M. M. New development in the theory of heat and mass transfer in solar stills, *Solar Energy*, 1995, vol. 55 (6), 527-535.

[8] Yang, S. A.; Li, G. C.; Yang, W. J. Thermodynamic optimization of free convective film condensation on a horizontal elliptical tube with variable wall temperature, *Int. J. Heat Mass Tran.*, 2007, vol. 50 (23-24), 4607-4613.

[9] Revankar, S. T.; Pollock, D. Laminar film condensation in a vertical tube in the presence of noncondensable gas, *App. Mathe. Modelling*, 2005, vol. 29, 341-359.

[10] Raach, H.; Mitrovic, J. Simulation of heat and mass transfer in a multi-effect distillation plant for sea water desalination, *Desalination*, 2007, vol. 204, 416-422.

[11] Nusselt, W. Die Oberflachenkondensation des Wasserdampfes, *Z. Ver. Deut. Ing.*, 1916, vol. 60, 541.

[12] Incropera, F. P.; DeWitt, D. P. *Fundamentals of Heat and Mass Transfer*, 4th ed., John Wiley & Sons, New York, 1996.

[13] Rohsenow, W. M. Heat transfer and temperature distribution in laminar film condensation, *Trans. ASME*, 1956, vol. 78, 1645.

[14] Sadasivan, P.; Lienhard, J. H. Sensible heat correction in laminar film boiling and condensation, *J. Heat Transfer*, 1987, vol. 109, 45.

[15] Sparrow, E. M.; Gregg, J. L. A boundary layer treatment of laminar film condensation, *J. Heat Transfer*, 1959, vol. 81, 13.

[16] Chen, M. M. An analytical study of laminar film condensation: Part 1-Flat Plate, *J. Heat Transfer*, 1961, vol. 83, 48.

[17] Dhir, V. K.; Lienhard, J. H. Laminar film condensation on plane and axissymmetric bodies in non-uniform gravity, *J. Heat Transfer*, 1971, vol. 93, 97.

[18] Ahsan, A.; Fukuhara, T. Condensation mass transfer in unsaturated humid air inside tubular solar still. *Annual J. of Hydraulic Engineering*, JSCE, 2009, vol. 53, 97-102.

[19] Ahsan, A.; Fukuhara, T. Condensation Mass Transfer in Unsaturated Humid Air inside Tubular Solar Still. *J. Hydroscience and Hydraulic Engineering*, JSCE, 2010, vol. 28(1), 31-42.

[20] Ahsan, A.; Islam K. M. S.; Fukuhara, T.; Ghazali, A. H. Experimental Study on Evaporation, Condensation and Production of a New Tubular Solar Still. *Desalination*, 2010, vol. 260 (1-3), 172-179.

[21] Ahsan, A.; Fukuhara, T. Mass and Heat Transfer Model of Tubular Solar Still. *Solar Energy*, 2010, vol. 84 (7), 1147-1156.

In: Water Condensation
Editor: Amimul Ahsan

ISBN: 978-1-61324-874-4

Chapter 5

EFFECT OF NONCONDENSABLE GASES ON WATER VAPOR CONDENSATION

Riad Benelmir*

University Henri Poincaré of Nancy, France

ABSTRACT

In this chapter we consider the effect of a noncondensable gas on water vapor condensation. Two geometries are investigated: condensation between two parallel plane plates and condensation in a finned tube heat exchanger (cold battery).

We start by a parametric study of condensation between two parallel plates. The model used for water vapor condensation is capable of predicting correctly the thermal performances with an error which does not exceed 7% of the measurements. The numerical results show that the mass fraction of air at the condensation surface increases with the mass fraction at inlet and there is a big concentration of the noncondensable gas at the liquid/mixture interface, which reduces the water vapor mass fraction and the condensate rate. The increase of inlet velocity leads to an increase of vapor mass fraction in the core of the mixture and the condensation surface and consequently an increase of the condensate rate. The increase of the molecular mass of the noncondensable gas leads to an increase of the water vapor mass fraction near the condensation surface(plate) and of the condensate rate. The increase in the inlet pressure causes a decrease of the air mass fraction near the wall and an increase of the water vapor mass fraction and the condensate rate.

*Email: riad.benelmir@lermab.uhp-nancy.fr.

In the case of a cold battery, the air mass fraction increases more rapidly near the fin than in the middle of the channel. The maximum value of the air mass fraction is found in the recirculation zone behind the tubes. The air mass fraction around the first tube(fin side) is smaller than its middle side value. The increase in inlet velocity causes a decrease of mass fraction around the tubes and increases the condensate rate and the heat transfer coefficient. Globally the condensate rate increases with velocity.

1. INTRODUCTION

Due to its industrial importance, researchers have been very active in the field of heat and mass transfer which occur during water vapor condensation in the presence of noncondensable gases. SIDDIQUE*ET al.* (1994) studied forced convection of a vapor in the presence of non-condensable gas inside a vertical tube. Tsay(1996) studied the effect of binary mixture condensation (air+water vapor) on heat and mass transfer in a turbulent flow with mixed convection. Dehbiand Guentay(1997) proposed a model to study the performance of a vertical tube condenser in the presence of a noncondensable gas. Chen *et al.* (1998) studied the effect of a noncondensable gas on film condensation along a vertical plane fin. Chin *et al.* (1998) present a boundary layer model for laminar condensation with mixed convection and in the presence of a noncondensable gas on an inclined plane. Tribes (1999) considered motionless condensate on the surface; he took account of the reduction in air flow section due to the condensate and associated conductivity to the condensate thickness so as to take into account its thermal resistance.

By going through all these research works, we find that even for small mass fractions of a noncondensable gas in the fluid, there is a big concentration of this gas at the liquid/mixture interface (Chen *et al.*(1998), Chin *et al.*(1998), Dehbiand Guentay(1997)). The increase in mass fraction of the noncondensable gas at inlet increases its value at the interface and decreases the heat transfer coefficient (Siddique*et al.* (1994)). This increase of the mass fraction at inlet also decreases the fin efficiency because it causes a decrease of heat transfer by condensation.

As to the effect of velocity and Reynolds number, the heat transfer coefficient and the sensible power transmitted at the wall increase with the inlet Reynolds number (Siddique*et al.* (1994), Tsay(1996), Dehbiand Guentay(1997)), one will also have a very strong water vapor condensation (Tsay(1996)). One can also have an increase of water vapor condensation by increasing relative humidity (Tsay(1996)).

In this chapter, we start by a presentation of the basic physical models which exist in the code used in numerical modeling and also the algorithms related to our work available in the code. Then we present the numerical analysis done on film condensation of a flow of humid air in presence of a noncondensable gas. This approach applies to several geometries (vertical plane plates and finned battery) for an incompressible fluid in a laminar flow and for stationary and transient regimes.

2. PHYSICAL MODELS

The conservation equations for a laminar incompressible flow are:

2.1. The Mass Conservation Equation

The mass conservation equation is:

$$\frac{\partial \rho}{\partial t}+\frac{\partial}{\partial x}(\rho u)+\frac{\partial}{\partial y}(\rho v)+\frac{\partial}{\partial z}(\rho w)=S_m \tag{1}$$

It is a general form of the continuity equation which is valid for compressible and incompressible flows. The source term S_m is an additional mass to the continuous phase of a second dispersed phase(e.g. case of vaporization of liquid droplets) or a user defined function(source) (UDF)[1].

For the case of a mixture, the code can predict the local mass fraction for each species ($W_{i'}$) by solving the convection diffusion equation for species i. This conservation equation takes the following general form:

$$\frac{\partial}{\partial t}(\rho W_{i'})+\frac{\partial}{\partial x}(\rho u W_{i'})+\frac{\partial}{\partial y}(\rho v W_{i'})+\frac{\partial}{\partial z}(\rho w W_{i'})=$$
$$-\frac{\partial}{\partial x}J_{i',x}-\frac{\partial}{\partial y}J_{i',y}-\frac{\partial}{\partial z}J_{i',z}+R_{i'}+S_{i'} \tag{2}$$

[1]*UDF*= User Defined Functions.

$R_{i'}$:mass rate of creation or depletion by reaction,

$S_{i'}$:rate of creation by addition of a dispersed phase or a user defined source function

$J_{i',i}$: diffusion flux due to concentration gradient between a specie i' and the mixture in direction i given by:

$$J_{i',x} = -\rho D_{i',m}\frac{\partial W_{i'}}{\partial x}$$
$$J_{i',y} = -\rho D_{i',m}\frac{\partial W_{i'}}{\partial y} \qquad (3)$$
$$J_{i',z} = -\rho D_{i',m}\frac{\partial W_{i'}}{\partial z}$$

$D_{i',m}$:diffusion coefficientof "i" in the mixture.

This equation is solved for n-1 species, where n is the total number of chemical species present in the system.

2.2. Momentum Conservation Equations

Momentum conservation equations in directions x, y and z are:

$$\frac{\partial}{\partial t}(\rho u)+\frac{\partial}{\partial x}(\rho uu)+\frac{\partial}{\partial y}(\rho uv)+\frac{\partial}{\partial z}(\rho uw)=-\frac{\partial p}{\partial x}+\frac{\partial \tau_{xx}}{\partial x}+\frac{\partial \tau_{xy}}{\partial y}+\frac{\partial \tau_{xz}}{\partial z}+\rho g_x+F_x$$
$$\frac{\partial}{\partial t}(\rho v)+\frac{\partial}{\partial x}(\rho vu)+\frac{\partial}{\partial y}(\rho vv)+\frac{\partial}{\partial z}(\rho vw)=-\frac{\partial p}{\partial y}+\frac{\partial \tau_{yx}}{\partial x}+\frac{\partial \tau_{yy}}{\partial y}+\frac{\partial \tau_{yz}}{\partial z}+\rho g_y+F_y$$
$$\frac{\partial}{\partial t}(\rho w)+\frac{\partial}{\partial x}(\rho wu)+\frac{\partial}{\partial y}(\rho wv)+\frac{\partial}{\partial z}(\rho ww)=-\frac{\partial p}{\partial z}+\frac{\partial \tau_{zx}}{\partial x}+\frac{\partial \tau_{zy}}{\partial y}+\frac{\partial \tau_{zz}}{\partial z}+\rho g_z+F_z \qquad (4)$$

Where:

p: static pressure,

τ_{ij} : stress tensor,

ρg_i : gravity force indirection i,

F_i: External force indirection i, which also contains source terms or user defined functions.

2.3. Energy Conservation Equation

The code used allows the analysis of heat transfer in fluid and solid regions. The energy equation in fluid regions written in terms of sensible enthalpy has the following form:

$$\frac{\partial}{\partial t}(\rho h)+\frac{\partial}{\partial x}[u(\rho h)]+\frac{\partial}{\partial y}[v(\rho h)]+\frac{\partial}{\partial z}[w(\rho h)]=$$
$$\frac{\partial}{\partial x}\left(k\frac{\partial T}{\partial x}-\sum_{j'}h_{j'}J_{j'}\right)+\frac{\partial}{\partial y}\left(k\frac{\partial T}{\partial y}-\sum_{j'}h_{j'}J_{j'}\right)+\frac{\partial}{\partial z}\left(k\frac{\partial T}{\partial z}-\sum_{j'}h_{j'}J_{j'}\right)+S_h \tag{5}$$

For a mixture of perfect gases in incompressible flow:

$$h=\sum_{j'}W_{j'}h_{j'}+\frac{p}{\rho} \tag{6}$$

$$h_{j'}=\int_{T_{ref}}^{T}c_{p,j'}dT \tag{7}$$

where:

h: sensible enthalpy.

$J_{j'}$: j' species diffusion flux,

$W_{j'}$:mass fraction of species j'

$T_{ref} = 298.15$ K

The two terms in the right hand side of equation (5) represent energy transfer due to conduction and to species diffusion. $S_h$is the eventual heat of reaction or all the other volumetric heat sources which can be defined by the UDF.

2.4. Capabilities of the Code Used

For an arbitrary scalar ϕ_k, the code solves one of the following three equations according to the method employed by the user:

- Taking into account the diffusiveflux: (without the convective flux):

$$\frac{\partial}{\partial t}(\rho\phi_k) - \frac{\partial}{\partial x_i}\left(\Gamma_k \frac{\partial \phi_k}{\partial x_i}\right) = S_{\phi_k}, k = 1,..., N \tag{8}$$

- Taking into account bothdiffusiveand convective fluxes:

$$\frac{\partial}{\partial t}(\rho\phi_k) + \frac{\partial}{\partial x_i}\left(\rho u_i \phi_k - \Gamma_k \frac{\partial \phi_k}{\partial x_i}\right) = S_{\phi_k}, k = 1,..., N \tag{9}$$

- Taking into account adiffusive flux and a particular convective flux, defined by the user via the UDF:

$$\frac{\partial}{\partial t}(\rho\phi_k) + \frac{\partial}{\partial x_i}\left(F_i \phi_k - \Gamma_k \frac{\partial \phi_k}{\partial x_i}\right) = S_{\phi_k}, k = 1,..., N \tag{10}$$

where Γ_k is the diffusioncoefficient and S_{ϕ_k} the source term. Both are supplied by the user for each of the N scalars.

Example:

For $\phi_k = W_{i'}$ in equation (8), one obtains (3) in stationary regime,

For $\phi_k = 1$ in equation (9), one obtains (1),

For $\phi_k = W_{i'}$ in equation (9), one obtains (2).

3. Modeling of Condensation

In the theoretical analysis of heat and mass transfer during the condensation of a vapor containing a noncondensable gas, two methods are generally used, the first is the analysis of the boundary layer and the second is the analogy between heat and mass transfer.

In the analysis of the boundary layer, the works consisted in numerical resolution of the equations governing the conservation of mass, momentum and heat in the boundary layers of the mixture(vapor/gas) and the liquid(condensate)

In the analogy between heat and mass transfer, the equations are based on heat balance at the interface liquid/mixture (vapor/gas), where the exchanged heat through the mixture boundary layer equals the heat transferred through the condensate film.

Heat transfer from the gaseous phase consists of a sensible power due to heat transfer between the mixture(vapor/gas) and the condensation surface, and a latent heat due to the condensation of the diffused vapor on the interface. Separate models for sensible and latent heats have been used and mass transfer coefficients have been obtained using the analogy with heat transfer.

In this chapter, we only solved the mixture (vapor/gas) boundary layer. In the first place, we did not take into account the condensate thickness. In order to take into account the effect of water vapor condensation, we used a UDF developed by BELL(2000) to model this water vapor condensation and its effect on heat and mass transfer. A UDF similar to that of BELL and based on the use of Fick's law, has also been developed by Russeil*et al.* (1997), Tribes *et al.* (1999), Jason *et al.* (2001), Russeil*et al.* (2000).

The call in FLUENT of the mass transfer phenomenon is achieved through a procedure named "Surface reaction" which serves to apply Fick's law on the condensation surface (Figure 1). The condensate rate will be:

$$\dot{m}'' = -\rho D \left.\frac{\partial W_v}{\partial n}\right|_{paroi} \tag{11}$$

The condensation model is applied according to the algorithm in Figure (2).

To ensure the model stability, it is necessary to use an underrelaxation coefficient, the condensate rate will be calculated by:

$$\dot{m}''_{new} = \dot{m}''_{old} + \beta_m(\dot{m}''_{new} - \dot{m}''_{old}) \tag{12}$$

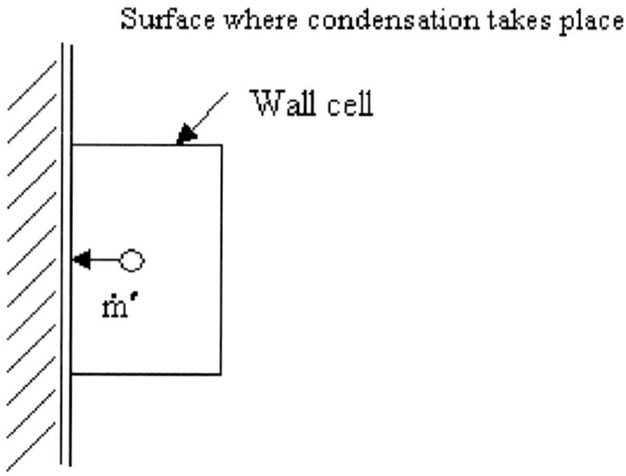

Figure 1.Condensation surface.

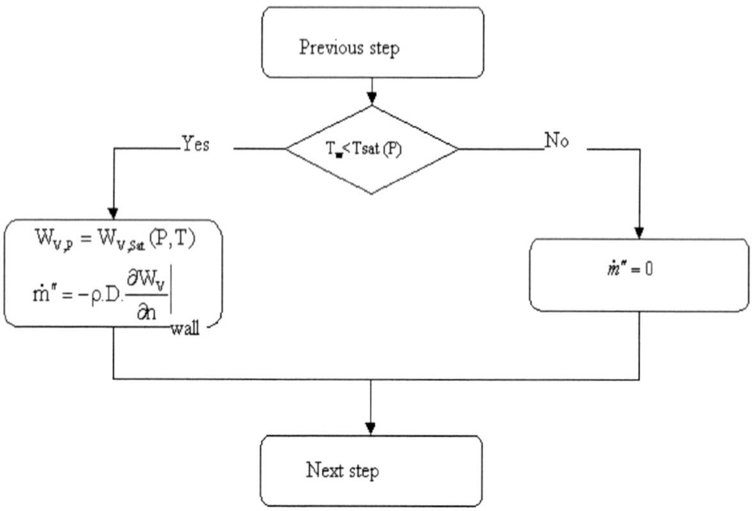

Figure 2. Algorithm applied tosurfacecondensation.

$\beta_m$is the underrelaxation factor,

$\dot{m}''_{new}$ is the rate calculated at the present iteration,

$\dot{m}''_{old}$ is the rate calculated at the previous iteration.

The global algorithm used in the calculations will be of the following form:

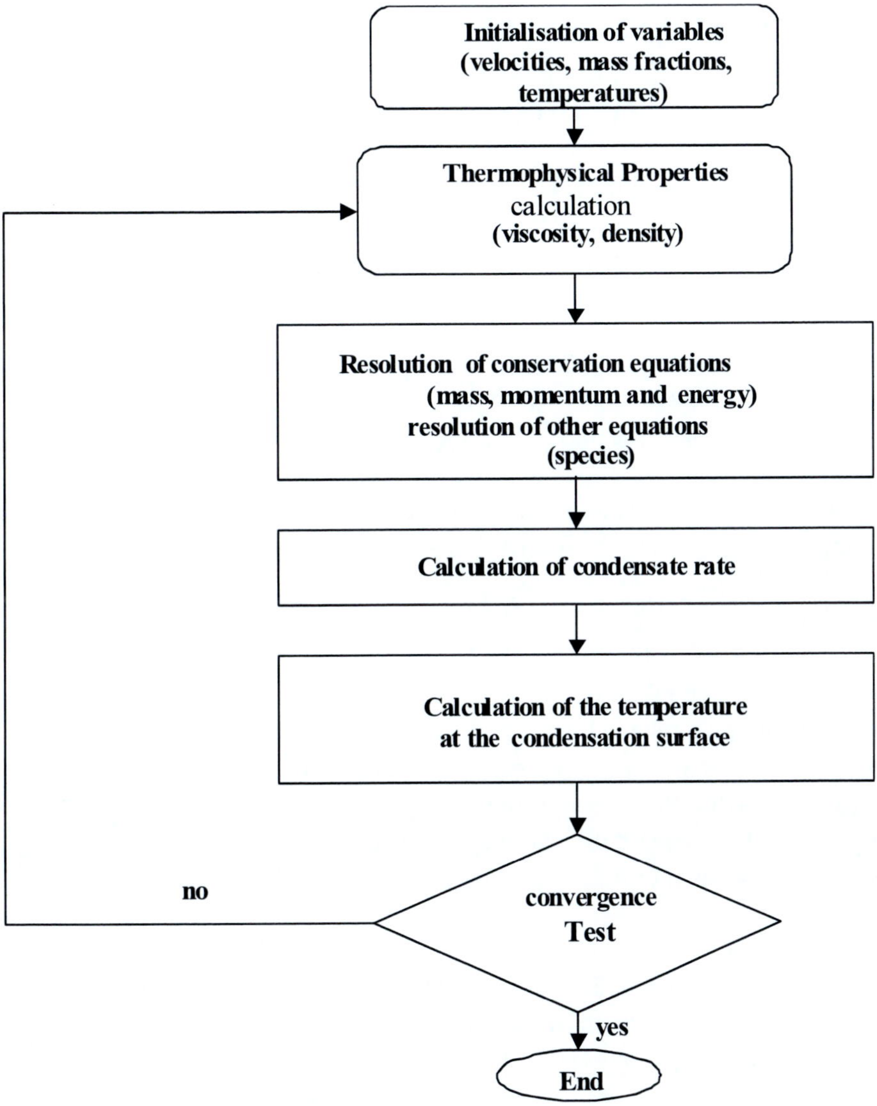

Figure 3.Global Algorithm used in the computations.

Based on this algorithm, many studies are presented in the following sections.

4. Numerical Analysis of Water Vapor ondensation in Presence of a Noncondensable Gas (Permanent Regime)

Souayed and Benelmir(2002, 2009) studied water vapor condensation in the presence of a noncondensable gas between two parallel plane plates (Figure 4). This case has been studied experimentally by Tribes (1999).

The geometric dimensions of the two plates are as follows:

- *L=370 mm*in thex direction.
- *l=367 mm*in they direction
- *e=8 mm*in thezdirection

In the numerical analysis in the permanent regime, the assumptions are the following:

- Laminar flow.
- Incompressible fluid.
- Newtonian fluid.
- The mixture assumed to be a perfect gas.
- condensation takes place at the interface, no volumetric condensation(formation of fog).
- Absence of viscou dissipation and radiation
- No dissolution of the noncondensable gas in the liquid phase.
- Negligible condensate thickness.

This model is represented by the conservation equations (1, 2, 4 and5). The computational domain is given by Figure 5.

The boundary conditions are:

Inlet.

For $x=0 T=T_\infty$, $u=U_\infty$, $w=0$, $W_v=W_{v\infty}$ (13)

planes (upper and lower).

$$\text{for } z=0 \text{ and } z=e=X_f+Y_f,\ \forall x u=0,\ w=0,\ \frac{\partial T}{\partial z}=0 \quad (14)$$

Plates (upper and lower).

$u=0$, $w=0$ and$T_w=278K$ (15)

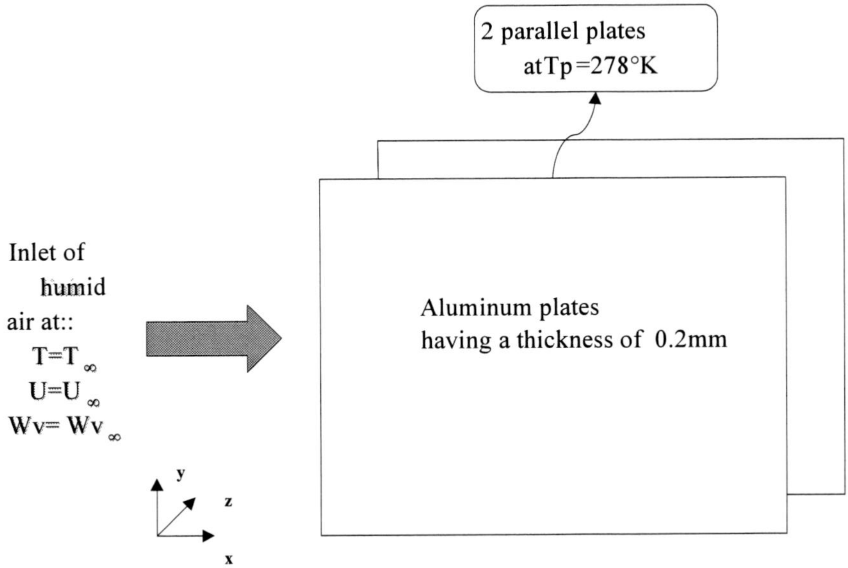

Figure 4. Two parallel plane plates.

Exit.

$$\text{For } x=(L+L_d),\ \forall z\ \frac{\partial u}{\partial x}=\frac{\partial w}{\partial x}=0 \quad \text{and} \quad \frac{\partial T}{\partial x}=0 \tag{16}$$

with Ld = 7.Dh=14e

Before validation of the model used, we give the values of the coefficients for humid air taken into account in the model:

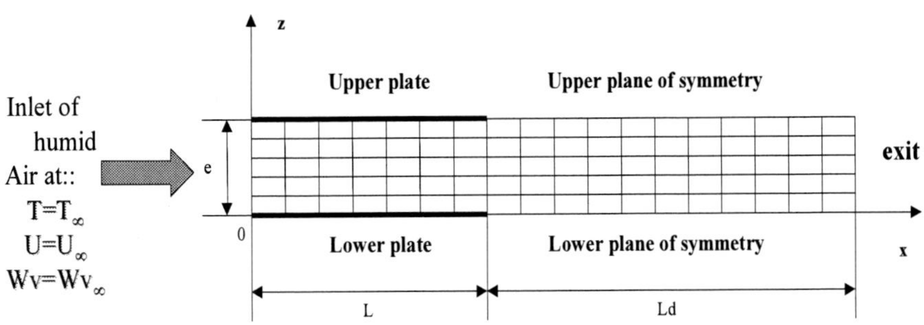

Figure 5.2D computational domain for the parallel plates.

$D_v = 2{,}065.10^{-5}\ m^2.s^{-1}$
$k = 0{,}0242\ W.m^{-1}.K^{-1}$
$Cp = 1013{,}484\ J.kg^{-1}.K^{-1}$
$\rho = 1{,}219\ kg.m^{-3}$
$\mu = 1{,}785.10\text{-}5\ kg.m^{-1}.s^{-1}$
$Pr = 0{,}747$

4.1. Validation of the Model Used

The exchange surfaces studied experimentally by Tribesare two parallel sheets of aluminum having a 0.2 mm thickness used usually in the manufacture of industrial heat exchangers fins. Each of the sheets is glued on a plate of aluminum having a thickness of 0.2 mm maintained on support plates cooled by circulation of glycoled water the temperature of which is regulated with a 3 way tap. The circulation of glycoled water is cocurrent with the humid air flow, in such a way as to minimize the temperature gradient along the aluminum sheet. The temperature of the heat exchange surface is measured by thermocouples placed between the sheets and the aluminum plates.

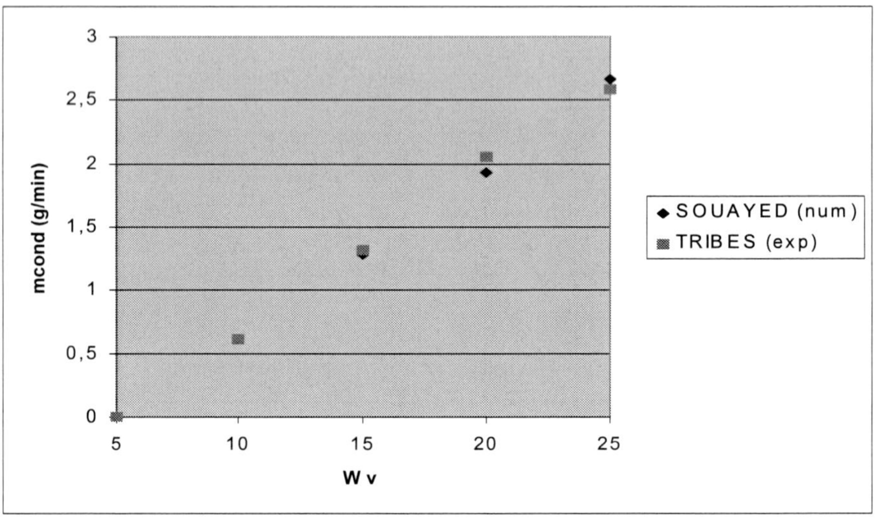

Figure 6.condensate rate as afunctionof the mass fraction of vapor $W_v$at the inlet (U_∞=1m/s, T_∞=308K et T_w=278K).

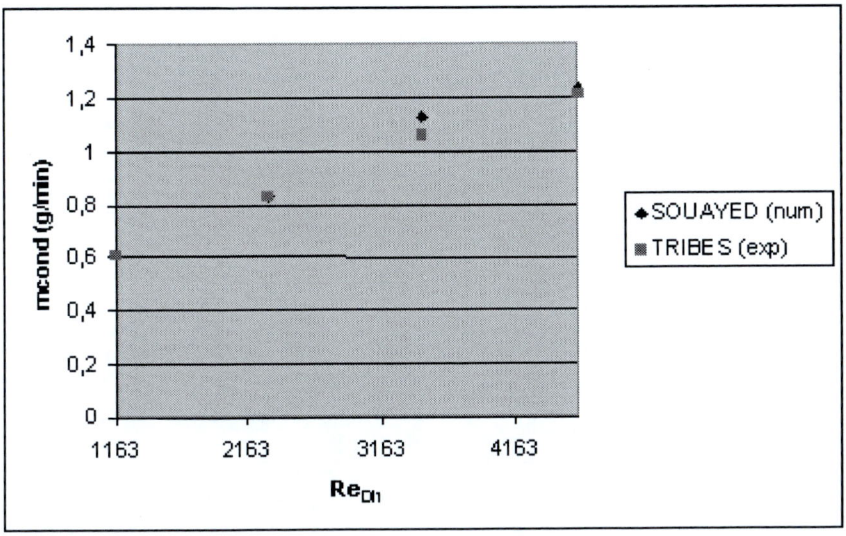

Figure 7.condensate rate as afunctionof Re_{Dh} ($W_{v\infty}$=10g/kg, T_{∞}=308K et T_w=278K).

The validation shows that the model computes the condensate rate with a relative error in comparison with the experimental analysis which does not exceed 6%(Figure 6) and 7% (Figure 7).

In the calculation of the sensible power, one notices that the relative error does not exceed 6% (Figure 8) and 10% (Figure 9). In Figure 9, one can see that the relative error between experimental and numerical results increases with the Reynolds number Re, which is normal since the model used is laminar and does not take into account heat dissipation which comesfrom turbulence which takes place starting from Re=2350 (for an increase of 66% in Re, the relative error increases by 98%)

In what follows, Souayed (2002)shows the effect of many parameters on water vapor condensation.

Effect of Air Mass Fraction at Inlet

The mass fraction of air(near the plate and in the mixture core) increases with its value at inlet (Figure 10 and Figure 11) due to the increase of the concentration of this gas at the interface liquid/mixture, which reinforces the resistance to water vapor transfer and lowers the mass fraction of water vapor near the plate (Siddique*et al.* (1994), Chen *et al.* (1998), CHIN *et al.* (1998)). On the other hand, the increase of the noncondensable gas mass fraction with x is due to the increase of the boundary layer thickness on the plate and the decrease of the concentration gradient of the noncondensable gas.

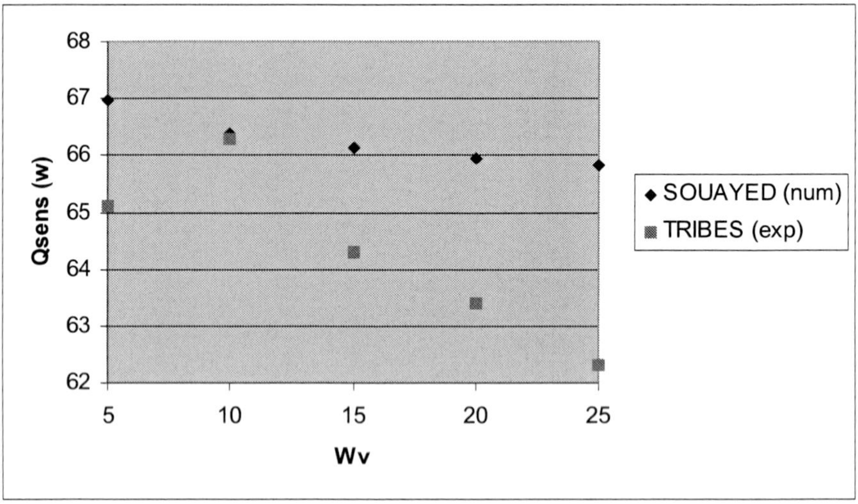

Figure 8. Variation of sensible power as a function of $W_v$at inlet (U_∞=1m/s, T_∞=308K and T_w=278K).

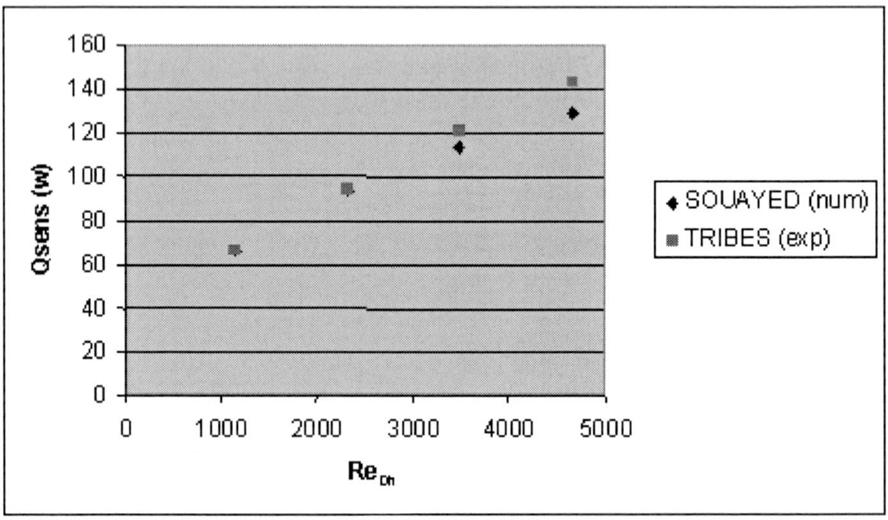

Figure 9. Variation of sensible power as a function ofRe ($W_{v\infty}$=10g/kg, T_∞=308K et T_w=278K).

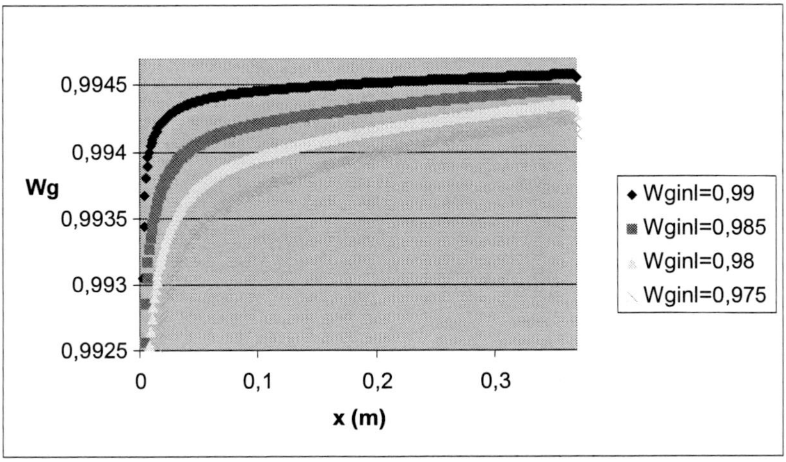

Figure 10. Evolution of the mass fraction of air near the plate as a function of the mass fraction of air at inlet (U_∞=1m/s, T_∞=308K and T_w=278K).

4.2. Parametric Study

The decrease of the water vapor mass fraction reduces the condensate rate (Figure 12).

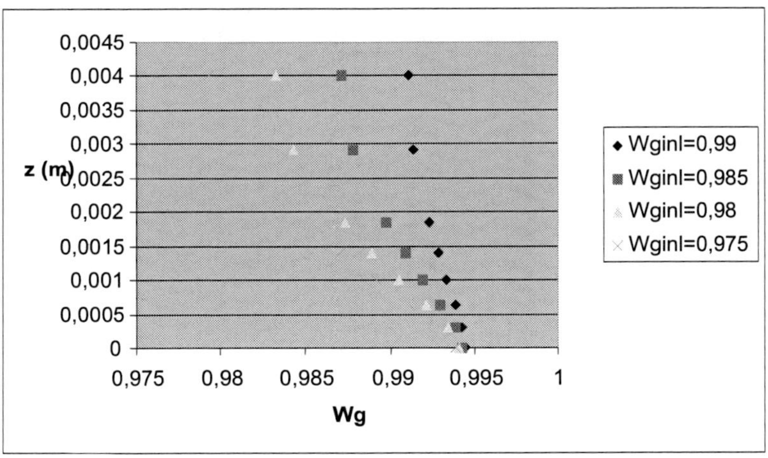

Figure 11. Variation of the profile of air mass fraction between the plate and the mixture core (U_∞=1m/s, x/L=0.27m, T_∞=308K andT_w=278K).

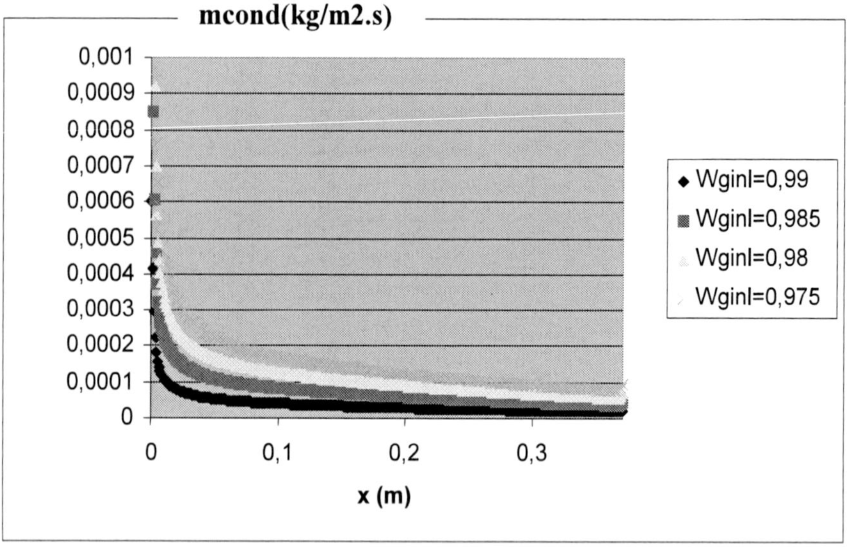

Figure 12. Evolution of the condensate rate near the plate (U_∞=1m/s, T_∞=308K and T_w=278K).

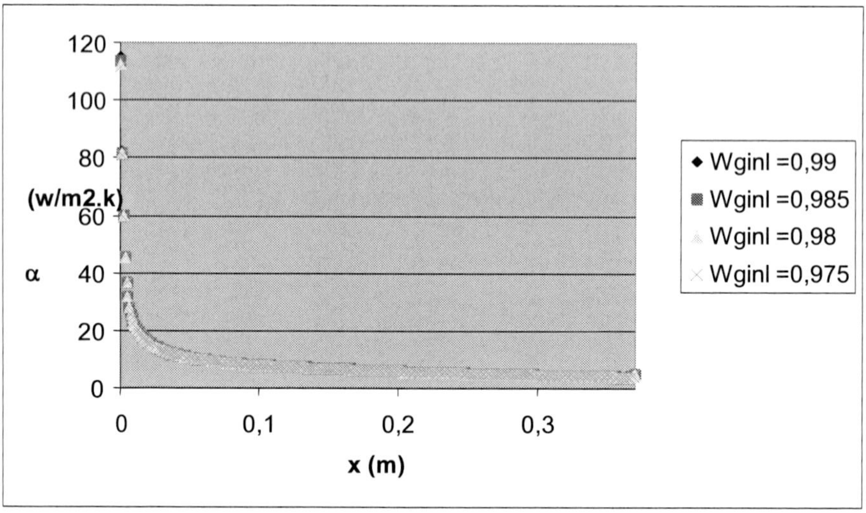

Figure 13. Evolution of the heat transfer coefficient along the plate(U_∞=1m/s, T_∞=308K and T_w=278K).

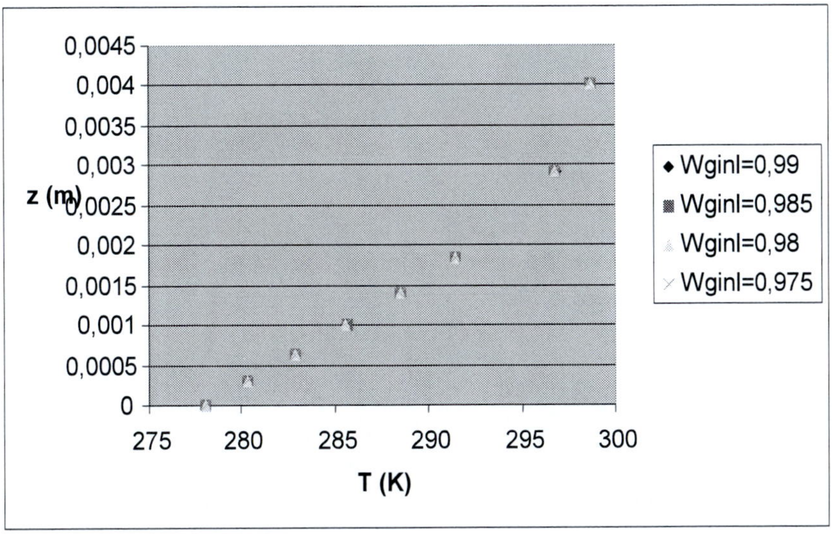

Figure 14. Temperature profile between the plate and the core of the mixture (U_∞=1m/s, x/L=0.27, T_∞=308K andT_w=278K).

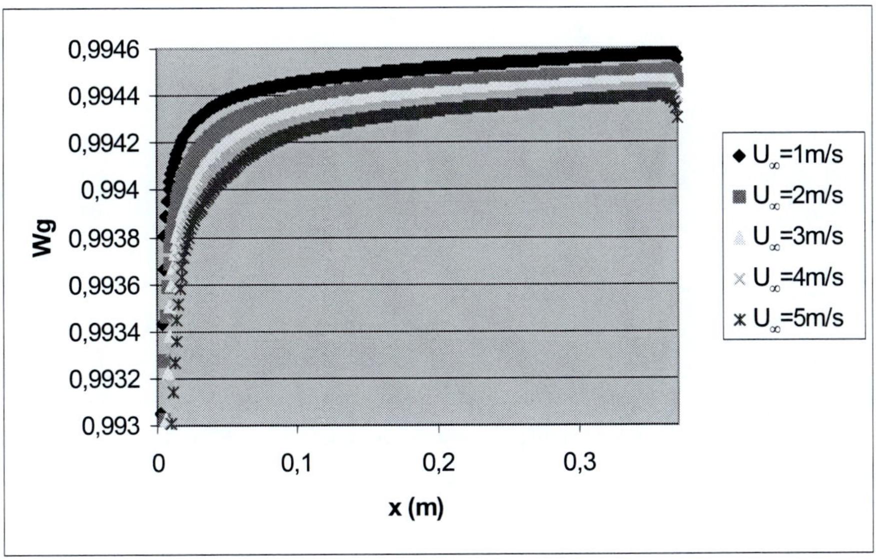

Figure 15. Evolution of the mass fraction of air near the plate as a function of the velocity at entry ($W_{v\infty}$=10g/kg, T_∞=308K and T_w=278K).

The heat transfer coefficient decreases slightly with the mass fraction of the noncondensable gas at inlet (Figure 13). This is due to the weak variation of the temperature fields near the plate and in the core of the mixture (Figure (14)) (Chen *et al.* (1998), Siddique*et al.* (1994), Dehbi et Guentay (1997)). The sudden drop at the entry of the two plates is due to the boundary layer development.

Effect of the Mixture Inlet Velocity

The mass fraction of air near the plate and in the core of the mixture decreases with the mixture velocity (air+water vapor) at inlet (Figure 15). This is due to the decrease of the thickness of the mass boundary layer(Figure 16), which increases the concentration gradient $\left(\frac{\partial C_a}{\partial z}\right)$ and leads to a decrease in the air mass fractionthereby causing an increase of water vapor concentration and consequently an increase of condensate rate(Figure 17). Dehbiand Guentay(1997) have shown that the effect of anoncondensablegas on the heat transfer coefficient is greater for low velocities than for high velocities, which implies that the effect of a noncondensable gas diminishes with the velocity.

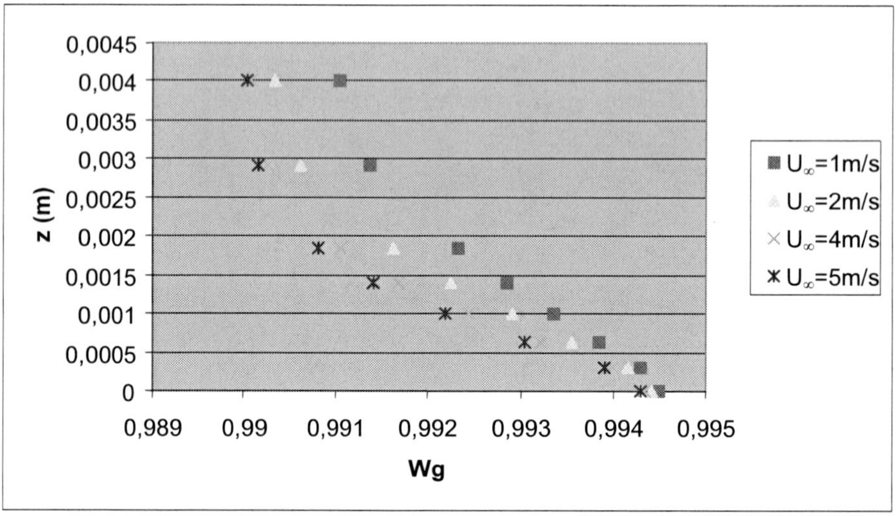

Figure 16. Masse fraction profile of air between the plate and the core of the mixture ($W_{v\infty}$=10g/kg, x/L=0.27, T_∞=308K and T_w=278K).

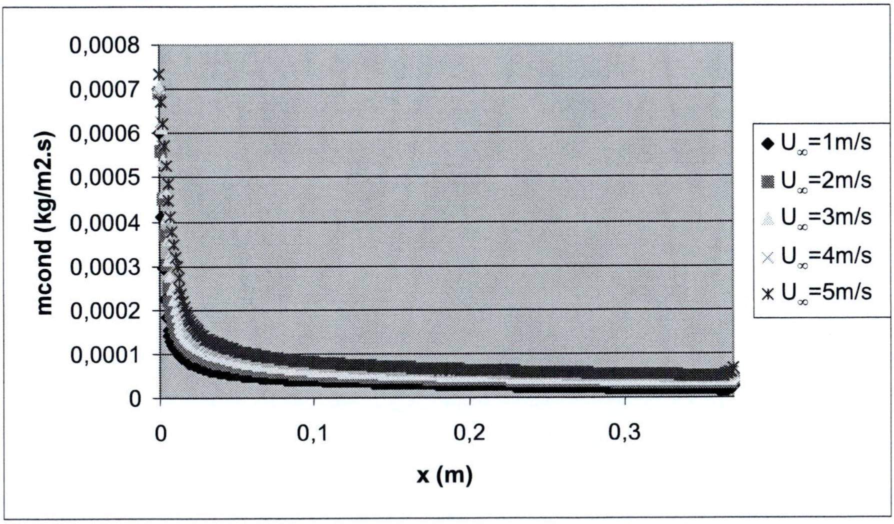

Figure 17. Evolution of the condensaterate near the plate as a function of the velocity at inlet ($W_{v\infty}$=10g/kg, T_{∞}=308K et T_w=278K).

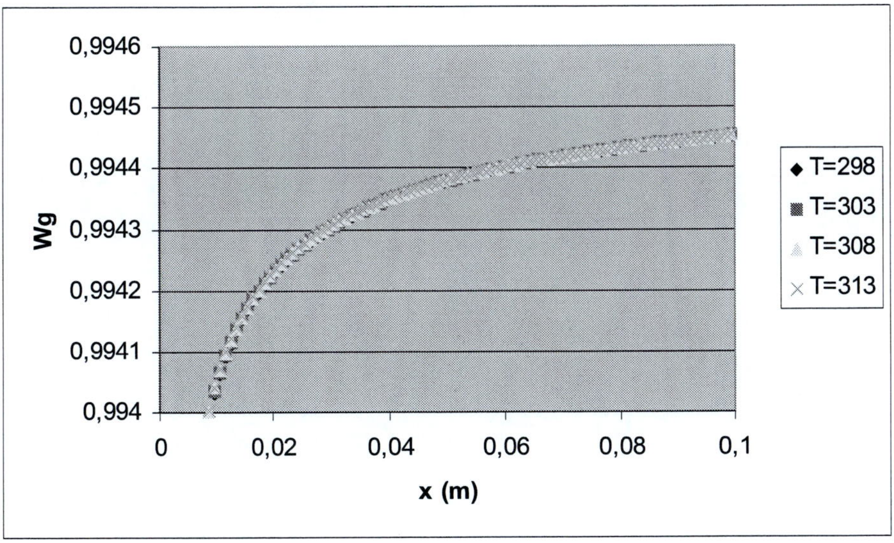

Figure 18. Evolution of the mass fraction of air near the plate as a function of the inlet temperature (U_{∞}=1m/s, $W_{v\infty}$=10g/kg and T_w=278K).

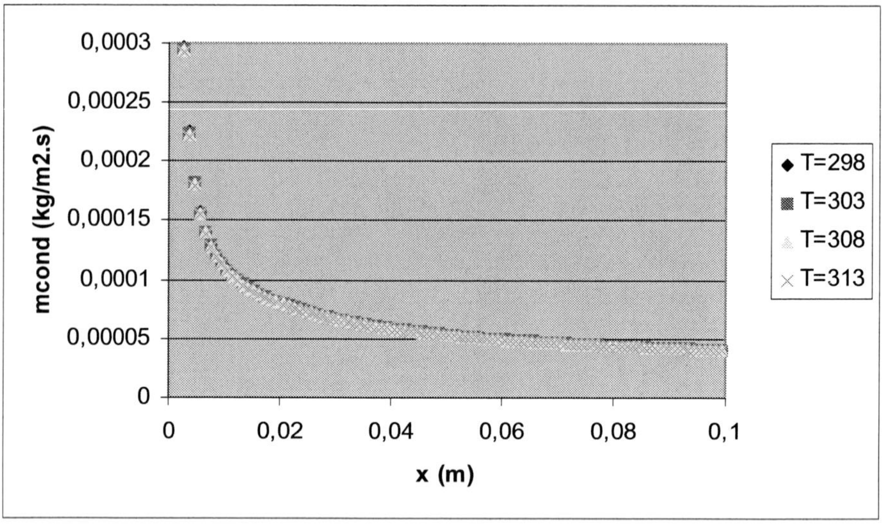

Figure 19. Evolution of the condensate rate near the plate as a function of the inlet temperature (U_∞=1m/s, Wv_∞=10g/kg and T_w=278K).

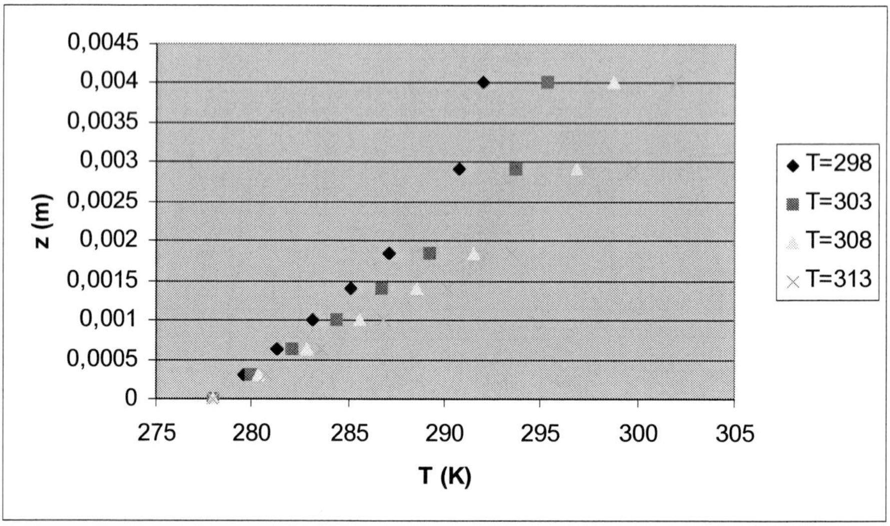

Figure 20. Variation of the temperature profile as a fonction de z (U_∞=1m/s, x/L=0.27, $Wv\infty$=10g/kg and T_w=278K).

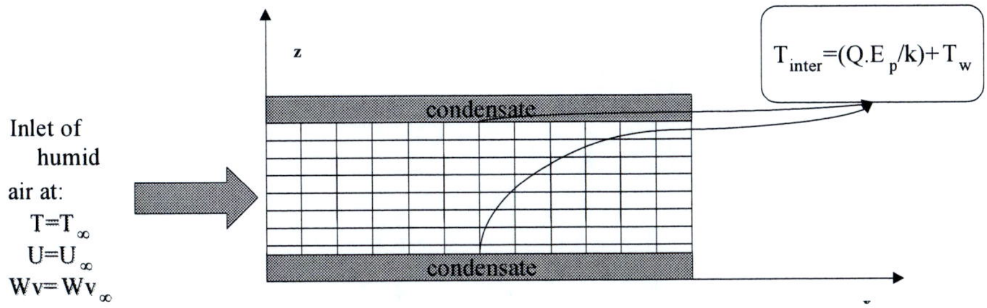

Figure 21.Computational domain when taking into account the condensate thickness.

Effect of the Mixture Inlet Temperature

By increasing the mixture inlet temperature, the mass fraction of air in the proximity of the plate varies slightly (Figure 18). Indeed, this temperature increase only leads to the variation of the thermal boundary layer and has a weak influence on the mass boundary layer(Dehbi et Guentay (1997)). This increase in entry temperature decreases slightly the condensate rate (Figure 19), because the difference between the mixture temperature and the plate temperature increases (Figure 20), which leads to an increase in the difference between water vapor partial pressure in the mixture and that at the condensation interface. The latter causes a decrease in vapor concentration and consequently the condensate rate. Effect of Condensate Thickness

In order to take into account the condensate thickness and to compute its effect on the power transmitted to the walls and on the mass fraction of the noncondensable gas, we calculated the temperature at the interface (gas/liquid) as a function of the condensate thickness and the condensation surface temperature(Figure 21). In this case the effective spacing between the plates will be reduced by twice the condensate thickness.

The results show that the mass fraction of air at the interface decreases with the condensate thickness(Figure 22). The increase of the condensate thickness leads to a decrease in the flow section and to the introduction of a supplementary thermal resistance. The decrease in flow section engenders more important mass fraction gradients of the noncondensable gas(or of the vapor) and also increases the velocity and consequently decreases the mass fraction of air at the interface(see § effect of mixture velocity). On the other hand, the introduction of a supplementary resistance decreases the power transmitted at the wall.

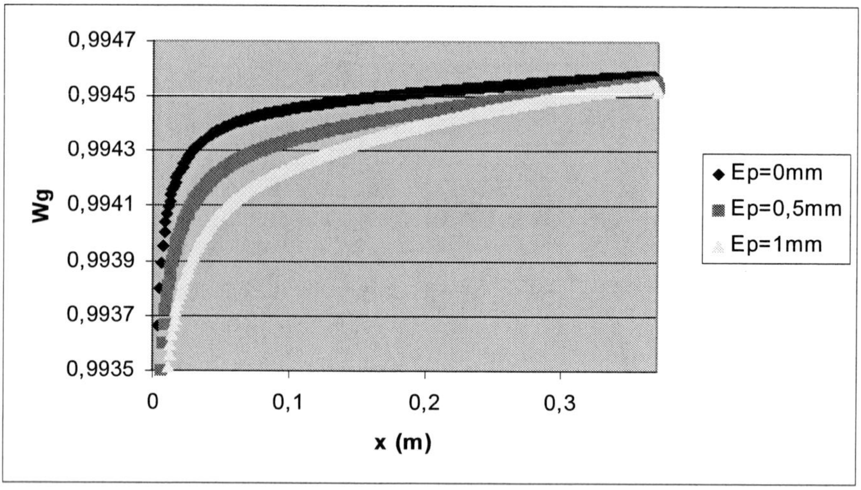

Figure 22. Evolution of the mass fraction of air along the interface(U_{∞}=1m/s, $W_{v\infty}$=10g/kg, T_{∞}=308K and T_w=278K).

The heat transfer coefficient decreases slightly with the thickness of condensate in particular at the end of the canal (Figure 23) due to the decrease of heat brought to the system since its volume decreases.

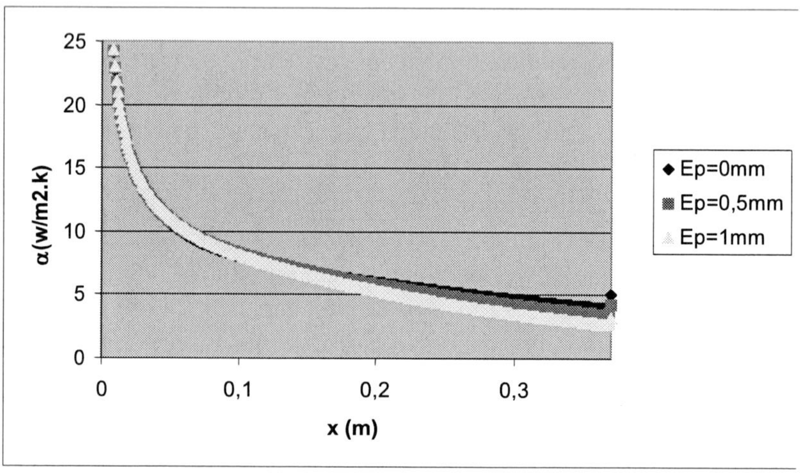

Figure 23. Variation of heat transfer coefficientalong the interface (U_{∞}=1m/s, $W_{v\infty}$=10g/kg, T_{∞}=308K and T_w=278K).

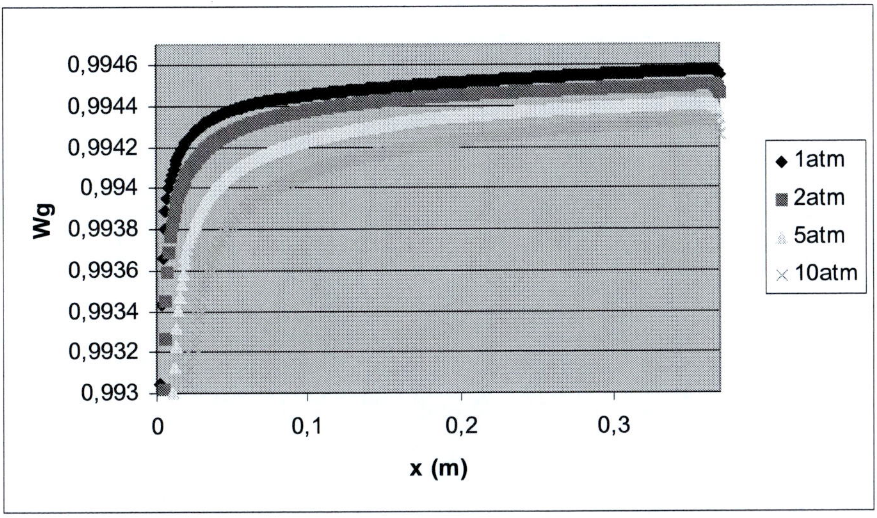

Figure 24. Evolution of mass fraction near the plate (U_∞=1m/s, $W_{v\infty}$=10g/kg, T_∞=308K and T_w=278K).

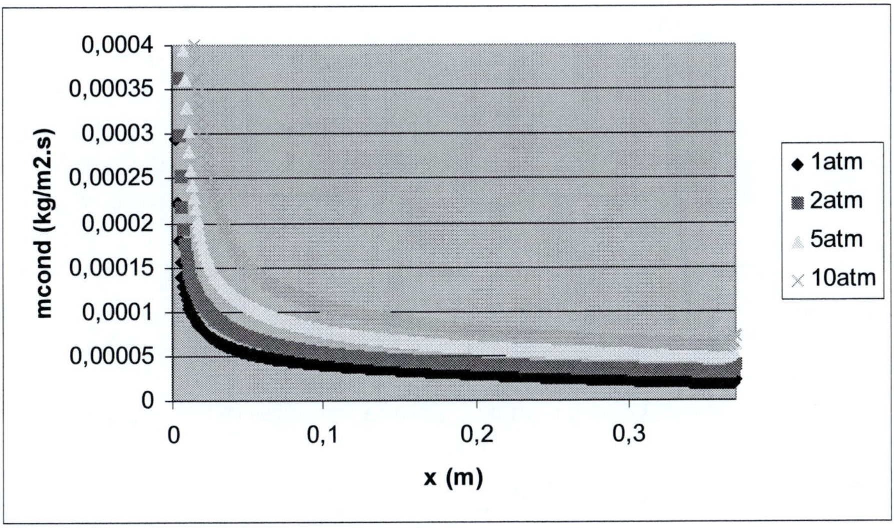

Figure 25. Evolution of condensate rate near the plate (U_∞=1m/s, $W_{v\infty}$=10g/kg, T_∞=308K et T_w=278K).

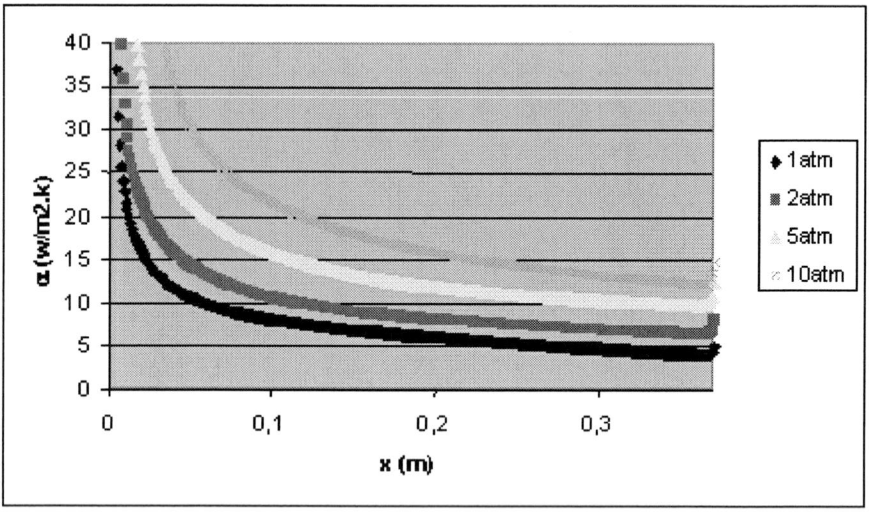

Figure 26. Evolution of heat transfer coefficient along the plate (U_∞=1m/s, $W_{v\infty}$=10g/kg, T_∞=308K and T_w=278K).

Effect of the Pressure

The mass fraction of air near the plate decreases with the pressure(Figure 24) due to the decrease of the diffusion coefficient, which increases the vapor mass fraction, the condensate rate (Figure 25) and the heat transfer coefficient increases with pressure (Figure 26). The decrease of diffusion coefficient diminishes with pressure, causes a decrease of the mass boundary layer thickness(increase of gradient of air concentration) and of the air mass fraction engendering an increase of water vapor concentration and consequently an increase of condensate rate(Figure 17).

Effect of Molecular Mass (Special Case)

Souayed(2002) substitutes to humid air other mixtures (Nitrogen/water vapor, Oxygen/water vapor, Hydrogen/water vaporand Atom O/water vapor). He finds that the mass fraction of water vapor near the plate increases with the molecular mass of the noncondensable gas[3]. (Figure 27), which causes a decrease of mass fraction of the *noncondensable gas near the plate, due to the reduction in the diffusion coefficient*(Siddique*et al.*(1994),Dehbiand Guentay(1997)).

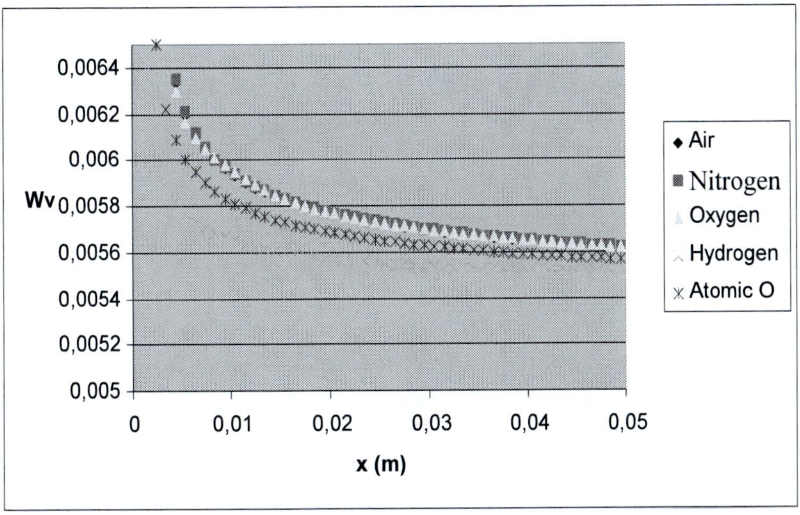

Figure 27. Evolution of vapor mass fraction near the plate (U_∞=1m/s, $W_{v\infty}$=10g/kg, T_∞=308K andT_w=278K).

By direct effect of the increase of water vapor mass fraction(near the plate), the condensate rate increases with the molecular mass. (Figure 28).

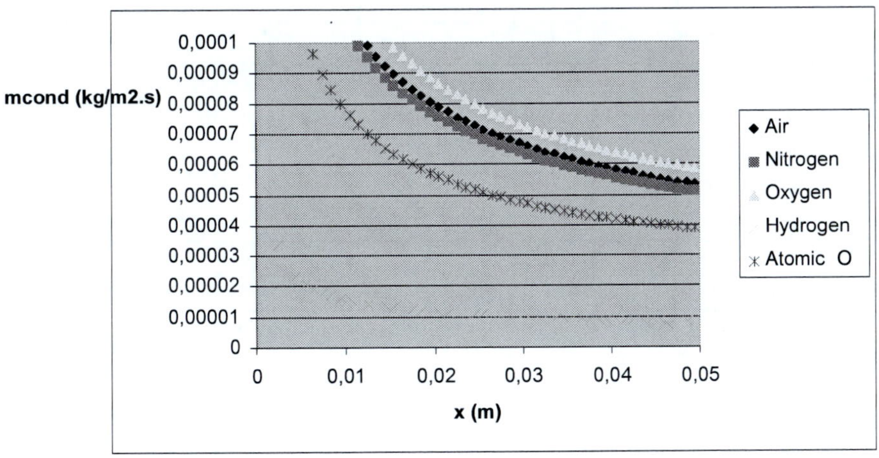

Figure 28. Evolution ofthe condensate rate near the plate (U_∞=1m/s, $W_{v\infty}$=10g/kg, T_∞=308K andT_w=278K).

Partial Conclusion

The mass fraction of a noncondensable gas at the interface increases with its value at inlet. This result has been confirmed by many authors (Siddiqueet *al.* (1994), Chen *et al.* (1998), Chin *et al.* (1998)) who showed that even for a small mass fraction of the noncondensable gas in the fluid, there is a big concentration of this gas at the interface liquid/mixture. The variation of the heat transfer coefficient is not very clear (visible) as we increase the air mass fraction because there is a weak variation of the temperature fields.

Some authors (Chen *et al.* (1998), Siddique*et al.*(1994)) show that the heat transfer coefficient and the fin efficiency decrease with the inlet mass fraction of the non condensable gas. (Chen *et al.* (1998), Dehbiand Guentay(1997)).

The increase of the inlet velocity, causes an increase of the temperature in the mixture core and the heat transfer coefficient (Siddique*et al.* (1994)), leading to an increase of heat transfer rate and the sensible power transmitted to the wall (Dehbiand Guentay(1997)). On the other hand, the increase of the velocity engenders an increase of the water vapor mass fraction in the mixture core and at the wall and consequently an increase of the condensate rate and the latent heat (Dehbiand Guentay(1997)).

Table 2.1.Dimensions of the exchanger

Length **of the exchanger**	0,262 m
Width of the exchanger	0,08 m
Hight of the exchanger	0,228 m
Number of tubes	45
Nombres of rows of tubes	5
Nombres of tubes per row	9
longitudinal pitch of the tubes	0,016 m
transversal pitch of the tubes	0,0258 m
Internal diameter of the tubes	8,7 mm
Diameter of the tubes	9,7 mm
Diameter + collar	10,1 mm
Number of fins	165
Thickness of fins	0,2 mm
Spacing between fins	1,6 mm
Fin density	625 fins/m (16 fins/in)
Fins in aluminium	

The introduction of condensate thickness in the computational domain affects the air mass fraction near the plate which diminishes with the condensate thickness. There is a weak decrease of heat transfer coefficient with condensate thickness.This decrease in sensible power with the condensate thickness is due to the decrease of the heat provided to the system.

The mass fraction of air near the plate decreases with pressure, which increases the condensate rate and the heat transfer coefficient. The molecular mass and the pressure have the same effect on the diffusion coefficient and an increase of the molecular mass and pressure provoke a decrease of the air mass fraction due to diffusion coefficient.

The mass fraction of water vapor near the plate increases with the molecular mass of the noncondensable gas. By direct effect of this increase in water vapor mass fraction, the condensate rate increases with the molecular mass (Dehbi et Guentay (1997)).

5. Numerical Analysis in Permanent Regime of Condensation in a Finned Tube Heat Exchanger in the Presence of a Noncondensable Gas

In this section, we present the results of the analysis of condensation in a heat exchangerin the presence of a noncondensable gas. It is a finned tube heat exchanger, the fins being planar and continuous, it contains 5 rows of tubes placed in triangular pitch. This exchanger was studied experimentally by Khalfi(1998). The dimensions of the exchanger are reported in the following table.

In this numerical study the following assumptions have been made:

- Laminar flow,
- Incompressible andNewtonian fluid
- mixture assumed to be a perfect gas,
- condensation at the interface (No volumetriccondensation),
- No viscous dissipation and radiation,
- No dissolution of the noncondensable gas in the liquid phase,
- Negligible condensate thikness.

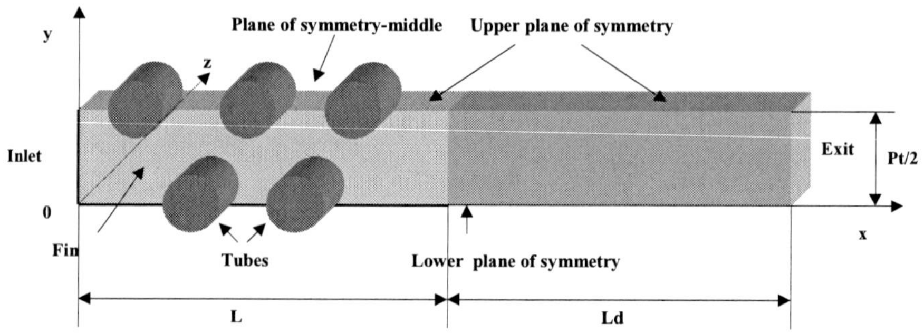

Figure 29. 3D computational domain of the heat exchanger.

This model is represented by the conservation equations (1, 2, 4 et 5). The computational domain is given in Figure 29.

The boundary conditions are as follows:

Inlet:

$$\text{For} x=0 T_\infty=308\text{K},\ u=U_\infty,\ v=0 \tag{17}$$

Horizontal planes of symmetry (upper and lower).

$$\text{For} y=0 \text{ and } y=P_t/2 v=0, \frac{\partial T}{\partial y}=0 \tag{18}$$

Verticalplane of symmetry (middle).

$$\text{For } z=X_f/2 w=0,\ \frac{\partial \text{T}}{\partial \text{z}}=0 \tag{19}$$

Fin and tubes.

$u=0$, $v=0$, $w=0$ and$T_w=278$K (20)

Exit.

$$\text{For } x=(L+L_d)\frac{\partial u}{\partial x}=\frac{\partial v}{\partial x}=\frac{\partial w}{\partial x}=0 \text{ and } \frac{\partial T}{\partial x}=0 \tag{21}$$

With $L_d = 7.D$

The numerical results show that the air mass fraction increases more rapidly near the fin than in the middle of the channel(Figures 30 to31). This result is due to the presence of the mass boundary layer of the noncondensable gas on the fin. Inside the mass boundary layer the air fraction varies more rapidly than in the middle of the domain and its value near the plate is higher than in the mixture core(see §effect of the masse fraction and Figure 11)

The maximum value of the air mass fraction is found in the recirculation region behind the tubes and near the fin. The effect of the velocity in this region will be less significant than in the other regions. This region constitutes a shelter for the noncondensable gas. Saboyaand Sparrow (1974) have noted the effect of the boundary layer developed in the front of the fin and the recirculation system which develops behind the tubes of the first row. They found that these two factors produce high mass transfer rates. For the second row of tubes, Saboyaand Sparrow found that only the recirculation system is responsible for the high mass transfer rates because there is no boundary layer development.

The air mass fraction around the first tube next to the fin is less than that near the flow core (Figure 32). This is due to the interruption of the mass boundary layer developed on the fin by the tubes. This value of the air mass fraction(near the fin) causes a condensate rate greater in this region than in the channel core (Figure 33).

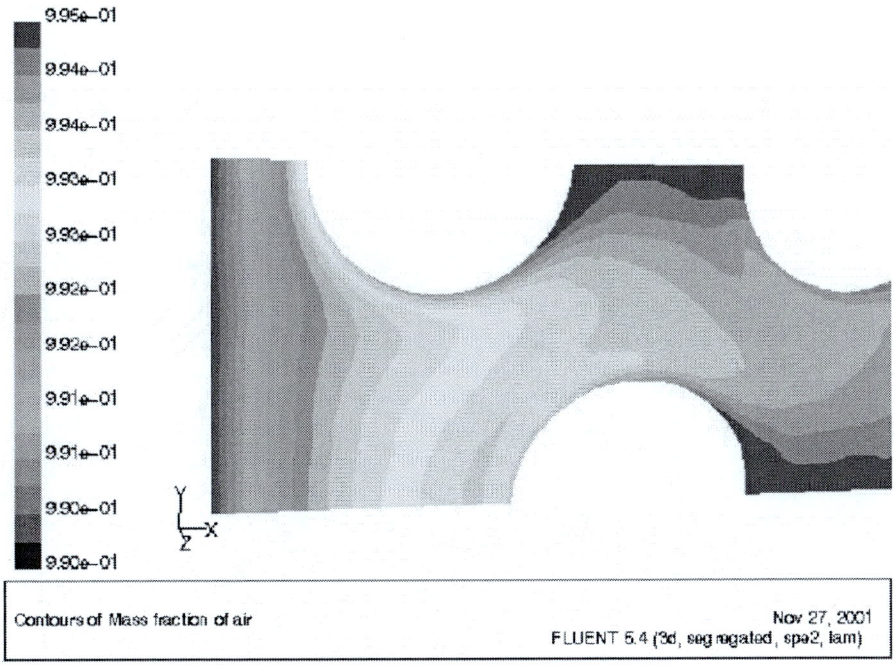

Figure 30. Air mass fraction near the fin.

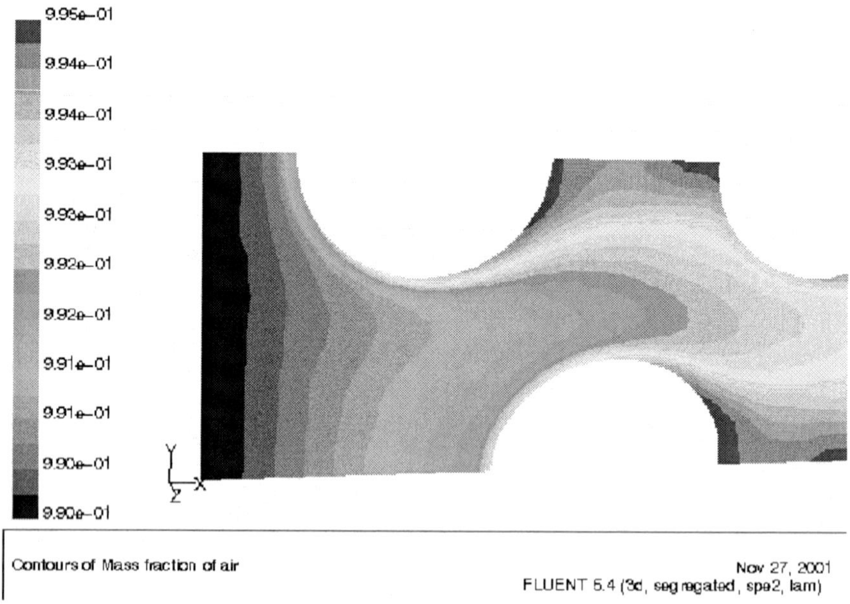

Figure 31. Air mass fraction in the channel core.

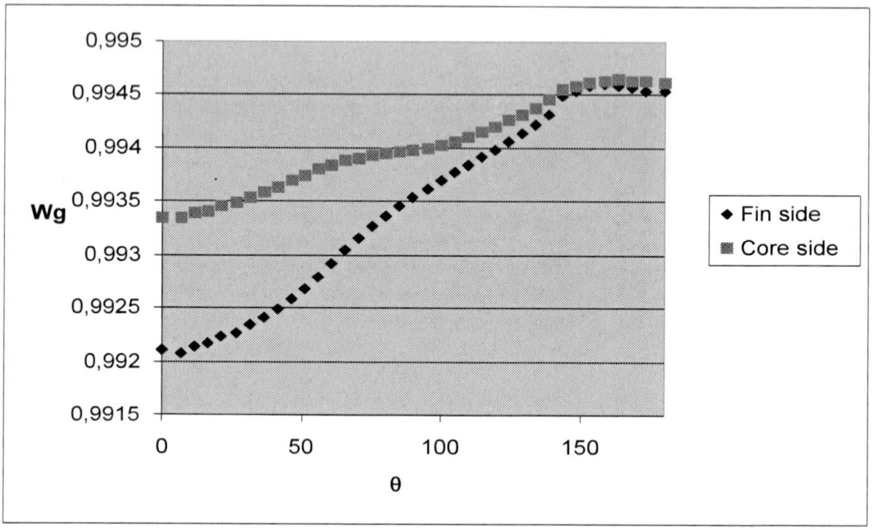

Figure 32. Comparison of the air mass fraction near the fin to that in the channel core.

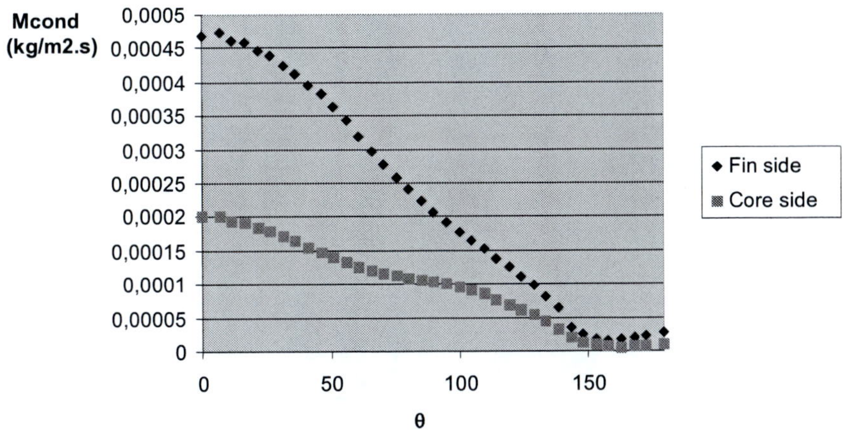

Figure 33. Evolution of the condensate rate around the first tube.

We notice that the maximum value of the heat transfer coefficient for the first and second tubes is in the front of these tubes and is about 55° away from the front for the other tubes (Figure 34) the dispositions of which can be seen in Figure 35.

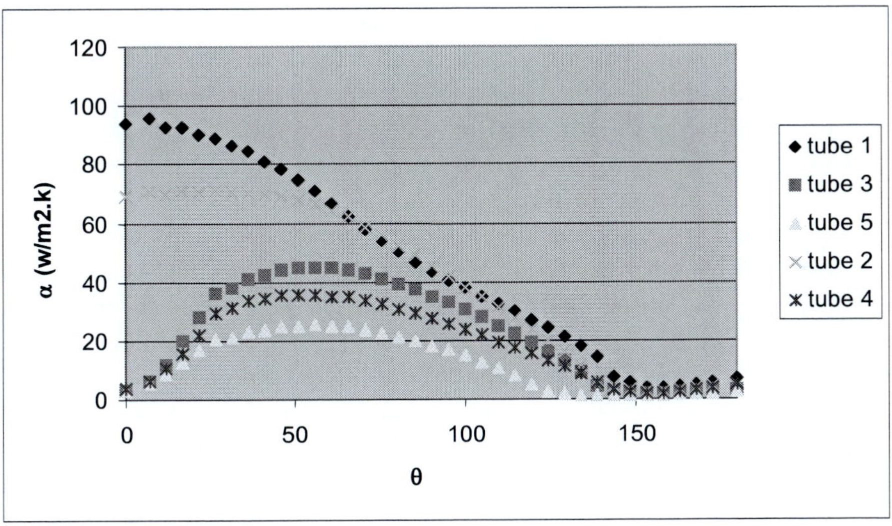

Figure 34. Evolution of heat transfer coefficient(fin side) around the5 tubes.

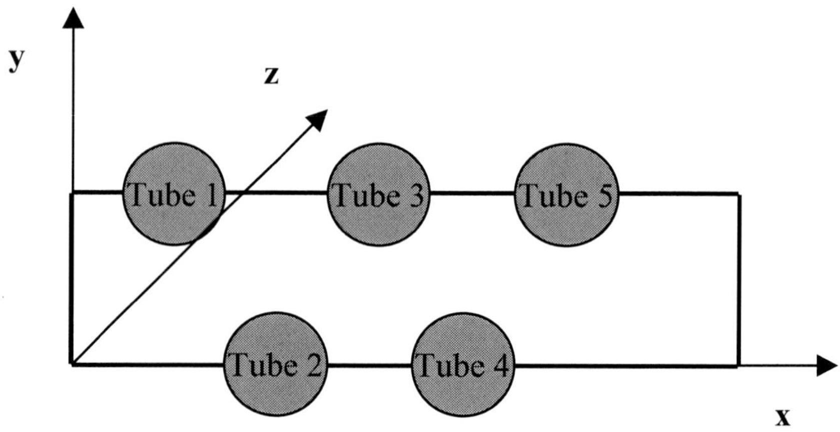

Figure 35. Disposition of the studied5 tubes.

This result is confirmed by many experimental and numerical studies (Rich (1973 and 1975), Wang*et al.*(1997), Fujii*et al.*(1984))which show that the heat transfer rate decreases with the number of tube rows and that the maximum value of the Nusselt number Nuis considerably bigger in the first row than for the other tube rows because there is no shadow effectin front of the tube.

The effect of the recirculation zone on the flow and heat transfer has been discussed in the works of many authors. Rosman*et al.*(1984) found that the recirculation region behind the tube is responsible for only 13 to 14% of all the heat transmitted by the fin (for 200<Re<1200). This result has been confirmed by Torikoshi*et al.*(1994) who show that this region contributes less to heat transfer than the frontal region of the tube where heat transfer is high(Figure 34 for the sensible heat)and that the formation of thisrecirculation zone depends strongly on fin spacing. On the other handWungand Chen (1989) noticed that because of the position of tube (1) in the first row, a recirculation zoneis formed in front and behind tube (3) in the second row (see previous figures 30 and 3)), provoking a decrease in heat transfer coefficient and the transmitted power. Kundu*et al.* (1991) observed that according to the value of Re and the geometry characteristics of the exchanger (in particular the tubes spacing), one can have a zone where there is no recirculation, another where their effects are negligible and a third zone where the effect of recirculation is considerable on the flow and heat transfer (see Figures 39 and (40). The decrease of heat transfer in this recirculation zone is confirmed in the works of Jang *et al.* (1996). Only Feibig*et al.*(1995) have spoken of fin heat transfer inversion of direction in this zone.

The noncondensable gas mass fraction increases with its value at inlet(Figure 36) leading to a decrease in condensate rate (Figure 37).

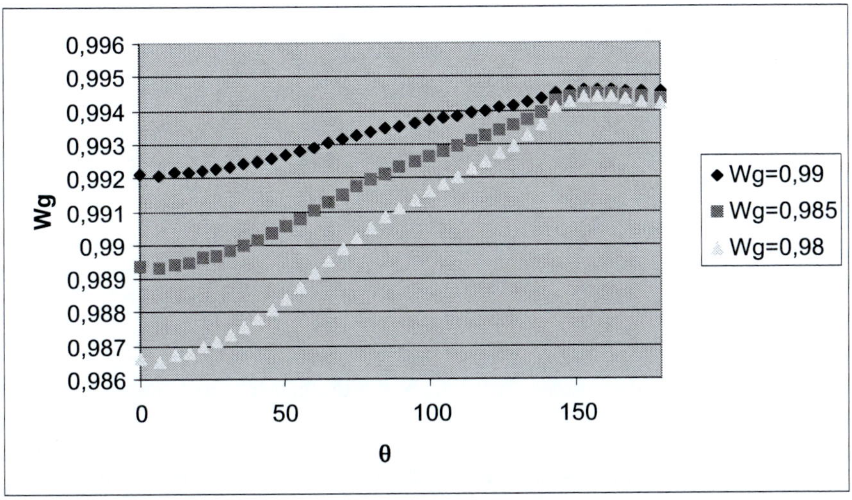

Figure 36. Evolution of the air mass fraction around the first tube as a function of Wg at inlet.

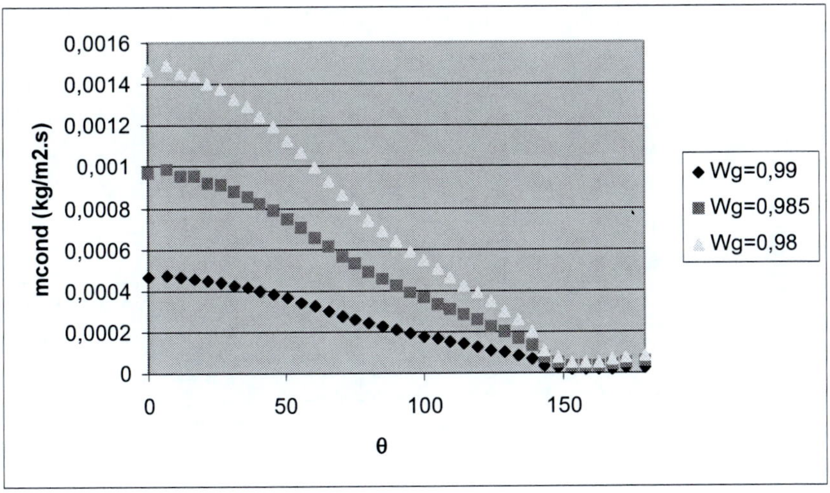

Figure 37. Evolution of the condensate rate around the first tube as a function of Wg at inlet.

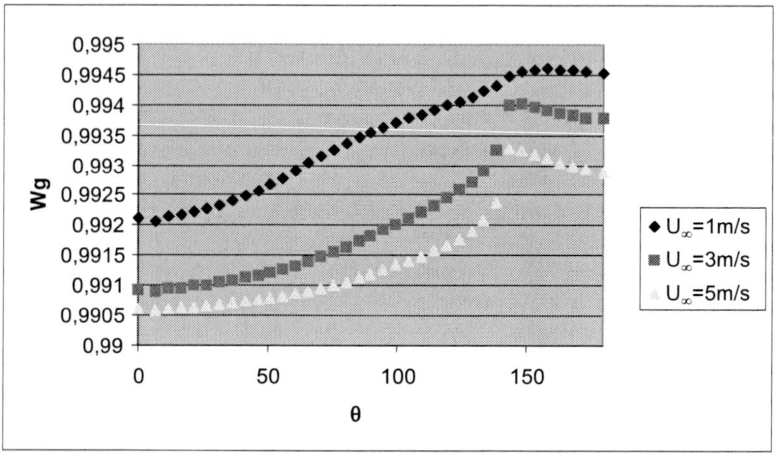

Figure 38. Evolution of the air mass fraction around the first tube as a function of the inlet velocity.

The non condensable gas mass fraction decreases with the inlet velocity (Figure 38) causing an increase in condensate rate(Figure 39). The maximum of this mass fraction of the non condensable gas is found behind the tubes (see also the recirculation zone in Figures 30 and 31). The increase in velocity improves the transfer and hence an increase of the heat transfer coefficient (Figure 40).

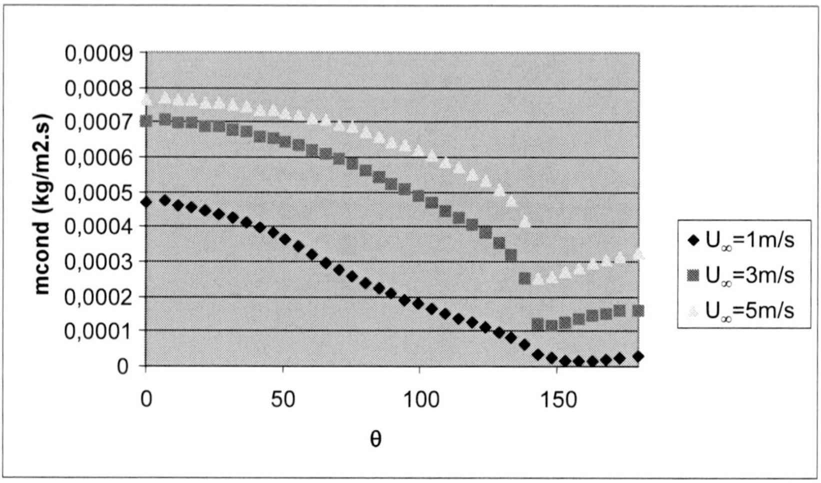

Figure 39. Evolution of the condensate rate around the first tube as a function of the inlet velocity.

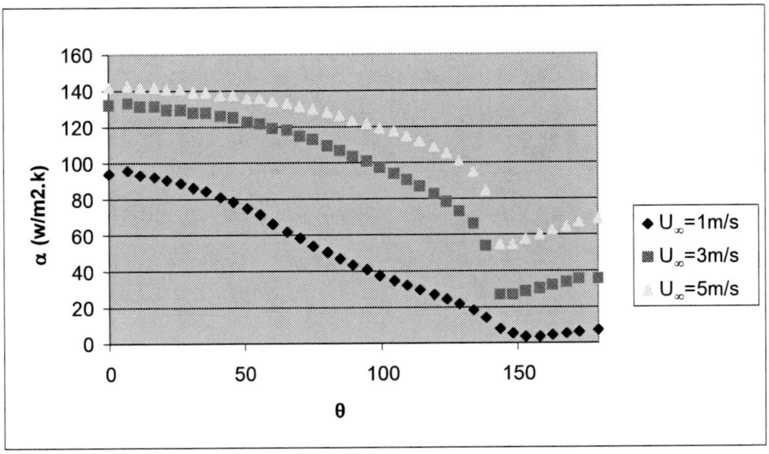

Figure 40. Evolution of the heat transfer coefficient around the first tube as a function of the inlet velocity.

In the global study of condensation in the finned tubes heat exchanger, the results show that the difference between the vapor mass fraction at inlet and that at the outlet of the heat exchanger is more important when the flow velocity is low (Figure 41) (Khalfi(1998)).

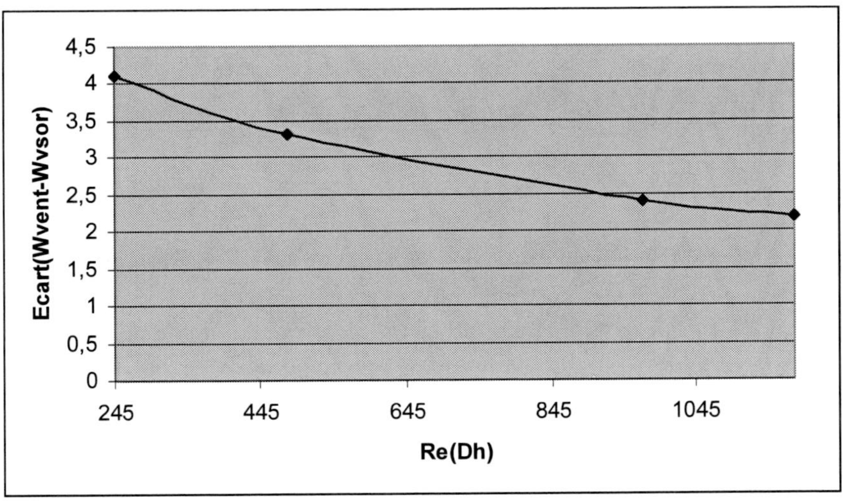

Figure 41. Variation of the difference between vapor mass fraction at inletand outlet as a function of ($W_{v\infty}$=10g/kg, T_∞=308K and T_w=278K).

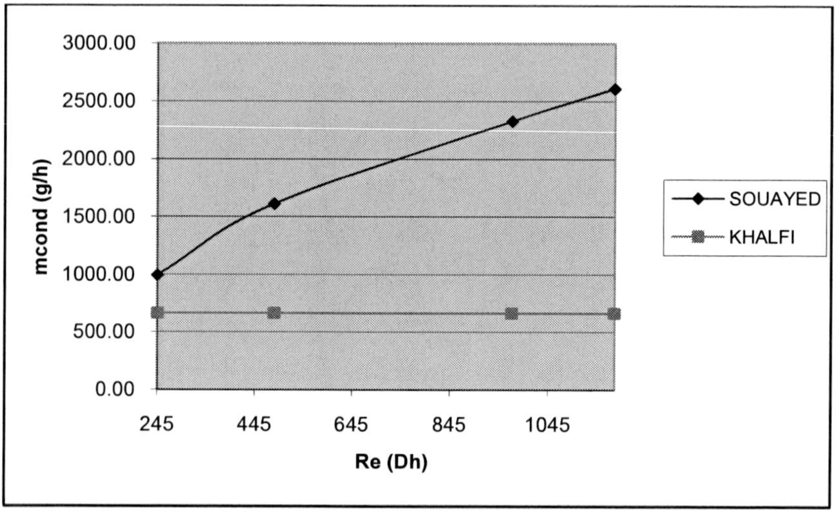

Figure 42. Evolution of the condensate rate recuperated by the exchanger as a function of Re ($W_{v\infty}$=10g/kg, T_{∞}=308K and T_w=278K).

The comparison between the present numerical results and the works of Khalfi(1998) show that there is a big difference between the condensate rate and the recuperated condensate rate due to a recuperation problem in Khalfi(1998) work, which necessitates a study of condensate droplet recuperation.

CONCLUSION

The results have shown that the air mass fraction at the interface increases with its inlet value and there is a big concentration of this gas at the interface liquid/mixture. This increase of noncondensable gas mass fraction at the interface will lower the vapor mass fraction and the condensate rate. The increase of the velocity at inlet leads to an increase of vapor mass fraction in the mixture core and at the interface and therefore toan increase in condensate rate. The increase of the molecular mass of the noncondensable gas causes an increase of water vapor mass fraction and condensate rate. The increase in inlet pressure decreases the air mass fraction near the plate and increases water vapor mass fraction and the condensate rate.

In the case of the finned tube heat exchanger, the air mass fraction increases more rapidly in the fin side than in the middle of the channel. The maximum value of the air mass fraction is found in the recirculation zone behind the tubes. The air mass fraction around the first tube (fin side) is smaller than that close to the middle(middle side) leading to a greater condensate rate in this zone than in the middle of the channel. The increase of inlet velocity diminishes the mass fraction around the tubes and increases the condensate rate and the heat transfer coefficient. Globally, the condensate rate increases with the inlet velocity..

The model presented in this chapter for water vapor condensation is capable of predicting correctly the thermal performances in the case of a heat exchanger and in the case of a simple geometry (two parallel plates) with a relative error which does not exceed 7% of the experimental results and allows us to study locally and globally the effect of certain parameters. We must note that only the mixtures with very high proportion of noncondensable gas are treated.

REFERENCES

[1] B. Bell (July 2000), UDF for condensation of steam from moist air, Fluent Inc.

[2] H. T. Chen, S. M. Chang, Z. Lan(1998), Effect on noncondensable gas on laminar film condensation along a vertical plate fin, *Int. J. Heat and Fluid Flow* 19, 374-381.

[3] Y. S. Chin, S. J. Ormiston, H. M. Soliman(1998), A two-phase boundary-layer model for laminar mixed-convection condensation with a noncondensable gas on inclined plates, *Heat and Mass Transfer* 34, 271-277.

[4] A. Dehbi, S. Guentay(1997), A model for the performance of a vertical tube condenser in the presence of noncondensable gases, *Nuclear Engineering and Design* 177 (1997) 41-52

[5] M. Fiebig*et al.* (1995), Conjugate heat transfer of a finned tube part A: heat transfer behavior and occurrence of heat transfer reversal. *Numerical heat transfer;*28:133-146.

[6] *Fluent 5 User's Guide* (July 1998), Fluent Inc..

[7] M. Fujii*et al.* (1984), A numerical analysis of laminar flow and heat transfer of air in an in-line tube bank. *Numerical heat transfer;*7: pp. 89-102.

[8] J.Y. Jang*etal.*(1996), Numerical and experimental studies of three dimensional plate fin and tube heat exchangers, *Inter. J. heat mass transfer,* Vol. 39, No. 14,pp. 3057-3066.

[9] J.M. Jason, S. Russeil, B. Baudoin(2001), Numerical modeling of the air side combined heat and mass transfer in a circular finned-tube heat exchanger under dry and wet conditions, *2nd International Conference on Computational Heat and Mass Transfer,* Rio de Janero, Brazil, October 226.

[10] M. S. Khalfi(1998), *Study of air humidity impact on the heat transfer coefficient in a cooling coil with and without condensation.*Doctoral Thesis (supervised by Prof. R. Benelmir), Nancy Henri Poincaré University, France.

[11] D. Kundu*et al.* (1991), Pressure and heat transfer in cross flow over cylinders between two parallel plates. *Numerical heat transfer;*19: pp. 345-360.

[12] S. V. Patankar(1980), *Numerical Heat Transfer and Fluid Flow,* Hemisphere Publishing Corporation, Washington.

[13] D.R. rich (1973), *The effect of fin spacing on the heat transfer anti friction performance of multi-rows, smooth plate fin-and-tube heat exchangers,* ASHRAE Annual Meeting, Louisville KY, June 24-28, pp. 137-145.

[14] D.R. rich (1975), The effect of the number of tube rows on the heat transfer performance of smooth plate fin-and-tube heat exchangers, *ASHRAE Transactions,* Part 1, Vol. 8 1, pp 307.

[15] E.C. rosman*et al.*(1984), -Performance of one and two row tube and plate fin heat exchangers, *ASME J. of heat transfer,* Vol. 106, pp. 627-632.

[16] S. Russeil, D. Marchal, B. Baudoin(1997), Air condensation in plate finned tube heat exchangers: a numerical model and an experimental set-up, *1st International Conference on Compact Heat Exchangers for the Process Industries,* Snowbird, USA, June 227, pp. 239-245.

[17] S. Russeil, C. Tribes, B. Baudoin (2000), *Etude de la condensation côté air dans les échangeurs de chaleur, Journée d'études GUT-SFT "Ebullition et Condensation",* Paris, 22 mars.

[18] F.E.M. saboya, E.M. sparrow (1976), transfer characteristics of two row plate fin and tube heat exchanger configurations, *Int. J. of heatmass transfer,* Vol. 19, pp. 41-49.

[19] M. Siddique, M. W. Golay,M. S. Kazimi(1994), *Theoretical modeling of forced convection condensation of steam in a vertical tube in the presence of noncondensable gas,* Heat Transfer and Fluid Flow 106 (May 1994) 20214.

[20] K. Torikoshi*et al.* (1994), Flow and heat transfer performance of a plate fin and tube heat exchanger (1st report: Effect of fin pitch), 10th Int. *Heat transfer Conf. Paper* 9-HE-16, pp. 411-416.

[21] C. Tribes(1999), *Etude du transfert de chaleur et de masse lors de la condensation de vapeur d'eau en présence d'incondensable,* Thèse de doctorat, Université de Valenciennes et du Hainaut-Cambraisis.

[22] C. Tribes, S. Russeil, B. Baudoin (1999), *Etude numérique et expérimentale de la condensation en présence d'air humide, 9èmes Journées Internationales de la Thermique,* Bruxelles, 6-8 Sept., Vol. 2, pp. 93-100.

[23] C.C. Wang *et al.*(1997), Performance of Plate Finned tube Heat Exchangers Under Dehumidifying Condition, *Journal of Heat Transfer,* Vol. 119, pp. 109-117.

[24] T.S. Wung, C.J. Chen (1989), Finite analytic solution of convective heat transfer for tube arrays in cross flow: part 1, flow field analysis, part 2, heat transfer analysis. *Journal of heat transfer;* 111:633-648.

[25] A. Souayed(2002), *Study of film condensation in a fin and tubes heat exchanger in presence of noncondensable gases. Doctoral Thesis* (supervised by Prof. R. Benelmir), Nancy Henri Poincaré University, France.

[26] R. Benelmir, S. Mokraoui, A. Souayed (2009), Numerical analysis of filmwise condensation in a plate fin-and-tube heat exchanger in presence of non condensable gas, *Heat and Mass Transfer Journal*, vol. 45, pp. 1561–1573.

•

INDEX

C

D

E

F

G

H

I

J

K

L

M

N

O

P

R

U

V

W

Y